Advanced Praise for *Rotating Machinery Reliability for Technicians and Engineers*

This book is a must-read for anyone involved in rotating equipment reliability, especially new technicians like myself. The author does an exceptional job of breaking down complicated engineering topics and explaining them in a way that is easy to understand.

Joe Nava, IPS Field Service Technician, Longview, WA

As an electric motor repair shop test stand technician, I have found Brook's book to be a valuable reference guide. The detailed index provides quick access to pertinent topics and case studies. The book is an excellent study guide for new technicians as well as experienced reliability professionals.

Joe Harakal, Field Service, HSE Coordinator,
IPS Rock Hill Service Center, Rock Hill, SC

I have known Ron for a long time—he has been a steadfast source of knowledge for many years. I am an experienced vibration analyst, certified CAT III. This book is hands-down the best I've ever seen, making difficult concepts easy to understand, and offering tremendous value to beginners and experienced techs/engineers alike. This is the book I wish I had when I started in the industry.

Blake Parker, Director of Shop Operations, Millington,
Hi-Speed Industrial Service, Millington, TN

I have read Ron's technical text and, as an engineer in a closely-related field, I found the text highly educational. The work was written in a much-needed "application-oriented" perspective, and would be extremely useful to field engineers. The book is wrapped around Ron's years of experience, which is what makes it priceless. The book provides a great background, and a chance to eliminate mistakes. Well done, Ron!

Peter Parazino, Field Engineer, DOD

This book has helped my team understand in more detail how to troubleshoot vibration issues in the field. In addition, my relationship with Ron has helped solve problems for our customers that our competition has not been able to do. A wonderful, helpful tool!

Erik C. Henderson, Account Manager, Integrated Power Services,
Charleston Service Center, North Charleston, SC

The strength of this book lies in the practical application of vibration analysis and the important subject of electrical testing. Ron does a nice job of explaining resonance testing and the practicalities of balancing, and has a good discussion of the Unbalance Tolerance Guide. I have been troubleshooting vibration problems for 40 years, and after reading this book, have no doubt that Ron has been on the front lines and is a true brother in arms. If someone is starting out in the business of troubleshooting vibration problems, they should read this book and absorb the nuggets of gold that Ron has extracted after many years of field experience.

Nelson L. Baxter, P.E., Category IV Vibration Specialist

Ron's book shows how analysts of 5 or 50 years in the field can help themselves in their approach to investigating potential issues in plant machinery. His collection of real-world problems captured from over 35 years of working in the field of reliability maintenance is an asset to all who follow the same path. I highly recommend the book to anyone who has a desire to learn and to better themselves in this craft, because I know that it's done wonders for me.

Elijah Brantley, Vibration Analyst, Thiele Kaolin Company, Sandersville, GA

Rotating Machinery
Reliability
for
Technicians and Engineers

Rotating Machinery Reliability
for
Technicians and Engineers

W. Ron Brook

INDUSTRIAL PRESS, INC.

Industrial Press, Inc.

32 Haviland Street, Suite 3
South Norwalk, Connecticut 06854
Phone: 203-956-5593

Toll-Free in USA: 888-528-7852
Email: info@industrialpress.com

Author: W. Ron Brook
Title: Rotating Machinery Reliability for Technicians and Engineers
Library of Congress Control Number: 2022946374

© by Industrial Press, Inc.
All rights reserved. Published in 2023.
Printed in the United States of America.

ISBN (print)	978-0-8311-3685-7
ISBN (ePUB)	978-0-8311-9622-6
ISBN (eMOBI)	978-0-8311-9623-3
ISBN (ePDF)	978-0-8311-9621-9

Publisher/Editorial Director: Judy Bass
Copy Editor: James Madru
Compositor: Patricia Wallenburg, TypeWriting
Proofreader: Mike McGee
Indexer: Arc Indexing, Inc.

books.industrialpress.com
ebooks.industrialpress.com
2 3 4 5 6 7 8 9 10

This book is dedicated to all of the engineers and technicians that have shared the results of their hard work. Their experiences add to the body of knowledge utilized by today's field personnel to analyze, diagnose, and remedy problems with rotating machinery of all types.

Field technicians, OEM engineers, and end users can struggle to solve machinery problems. There aren't many issues encountered that haven't been identified and solved previously. That is the duty of every person involved in this field. Document your experiences so that future field test personnel don't have to "reinvent the wheel."

Finally, to all managers of field service technicians: Give them time to learn.

The book has discussions on modal analysis and other technical items that are not possible to display in a book. The author has provided a website for the reader to download these articles, AVI files, and more at www.rotatingmachinereliability.com. They are indicated with one of these three icons:

Contents

Foreword

I first met Ron Brook at Reliance Electric Corporation in 1993. We had a bearing EDM issue with a specific Hermetic motor application and needed some assistance in the field. Ron worked in the services division of Reliance Electric and was highly recommended by another engineer in our group, William Miller. Ron determined that circulating current in the shaft from common mode voltage stresses induced by the VFD was the root cause of our bearing fluting. We were unable to replicate this until the motor was operating in an oil and freon environment.

This was the start of a great friendship that spanned the next 27 years and continues today as Ron is an Engineer Emeritus in Fleet Engineering for Integrated Power Services providing consultative services to our Field Services Team. *We never cease to be amazed by Ron.* Typically, we reach out to him for assistance, and before we can finish explaining our findings, he has the next steps or a solution in mind.

Ron is a highly-regarded rotating equipment technologist who has tackled many of our industries most challenging application issues. Technical Associates of Charlotte classified Ron as a Category 5 Vibration Analyst, which includes more than 25 years of experience in the field, plus a working knowledge of rotor dynamics modeling, modal analysis, and finite element analysis. The Category 5 analysts must also have sufficient experience and subject matter knowledge so they can be called for expert testimony during litigation. Ron paved a path for us and left an indelible mark on our industry.

Ron's book is the culmination of his 45 years of service to our industry. It provides invaluable insights to anyone who services and maintains rotating equipment. I would encourage all engineers and technicians who design, troubleshoot, or maintain rotating assets to read this book and absorb the concepts and diagnostic tools. It is a tremendous resource that provides a testament and legacy to the author.

Tom Reid
Senior VP Engineering and Nuclear Services
Integrated Power Services

Acknowledgments

I have met thousands of people over my 45-year career. Some were excellent teachers; others were vendors whom I kept for my entire career just based on how they treated their customers. I've been blessed with customers that were eager to learn about their machinery problems. I've stood in front of a room full of people wanting to learn about the latest frequency analyzer and how it could benefit their company, worked with those who stayed and helped even though quitting time had come and gone, and managers who procured the equipment needed to perform the tasks at hand. The work is still interesting and that is why I still consult in retirement. This list of names will not include all those who helped and provided inspiration for this book, but I've tried to spotlight the most influential of those folks. They comprise the following:

Roy Hench, who managed to get me through high school physics class; Robert Leon, who introduced his young field service crew to spectrum analyzers; Ralph Buscarello, my first formal vibration instructor; Charles Andrews, founder of DynaVibes Corporation; and Peter Phillips, the earliest coworker who tackled field problems.

Others include Patrick Link, the author of the first meaningful paper on identifying and solving electric discharge machining damage in bearings; Henry Bickel, Dick Rothchild, Reinhold Vogel, the brain trust of Nicolet Scientific, who believed a small crew of field-savvy technicians could add to the exposure of their product to field test engineers and technicians, when they purchased DynaVibes Corporation in 1980; George "Fox" Lang, Joe Deery, Randolph Perry, and Jack Heeg, who were all key to the success of the FFT; Dr. Donald Houser, of Ohio State University, who founded a museum dedicated to the FFT; Gerald Zobrist, President of Zonic Corporation, who believed a field technician could sell multi-channel signal processors and learn modal analysis; and Bill Owens, my first customer at Zonic Corporation.

There's also Jim Lally, President of PCB Corporation, who only wanted to hear his customer say, "I am a thoroughly satisfied customer" (I was for more than 40 years); William H. Miller, inventor and engineer, who taught me rotor dynamics modeling; Welton (Doc) Blessing, my "go-to" answer man for anything involving motors or generators; Dr. William B. Fagerstrom, for his dedication to the reliability field; and Norm Stalker, for believing my results regarding any field warranty issue.

I'd also like to thank Jim Berry, for his dedication to detail in describing vibration phenomena; James D'Angelo, for always being available to answer my questions; Tom Flynn, for being the kind of manager any worker would appreciate; Scott Crickenberger and Tom Watson of Daiken/McQuay, for their 20 plus years of relying on me for the "tough" problems; and Peter Paranzino, for introducing me to LabVIEW. Chapter 10 is dedicated to him.

To Tom Reid, without whose encouragement this book would still be in "bits and pieces" lying around my office and stuck in files inside old hard drives and servers. I will always proudly look back on his support with those "difficult" customers. It always helps to know that the VP of Engineering has your back.

The person who deserves the most credit, however, is my wife, Joan, for her patience and understanding through my last 15 years of work and the 5 long years it took to complete this book.

CHAPTER 1

Converting Dynamic Vibration into Electrical Signals

ACCELEROMETERS

We are surrounded by these devices without being aware of them. They keep display images in an upright position on screens of digital cameras and tablets. They stabilize drones in flight. High-frequency recording of biaxial and triaxial acceleration in biological applications are used to discriminate the behavioral patterns of animals. Their use in machinery health monitoring is the main focus in this book.

The workhorse of the industry is the piezoelectric accelerometer (Figure 1.1). The accelerometer has at its core a piezoelectric crystal. A crystal, when compressed, will yield a charge. The higher the force, the higher the charge output. The crystal is mounted on a steel base. A mass is placed on top of the crystal. It is this mass that reacts to the vibration input at the base.

When the steel base is mounted on a machine or structure, the vibration imparted to it causes the mass on top of the crystal to stress the crystal. This cyclic stress on the crystal causes a charge output proportional to the acceleration of the base. Charge-sensitive signals are affected by heat, vibration, stress, and so on. Early accelerometers did not have built-in electronic signal conditioners. The cable therefore would need to be anchored to prevent movement for its entire run, and heat and cold changes along the cable length had to be avoided at all costs. The output was

Figure 1.1   Typical industrial-grade piezoelectric accelerometer.
(Image courtesy of PCB Piezotronics, Inc.)

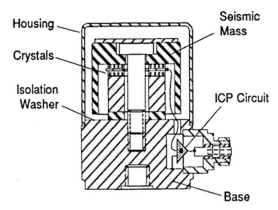

Figure 1.2   Internal construction of a typical piezoelectric accelerometer.
(Image courtesy of PCB Piezotronics, Inc.)

then plugged into a charge amplifier. This expensive piece of instrumentation yields a voltage output (Figure 1.3). The voltage output then can be analyzed. For this reason, the original devices were primarily used in aerospace and military applications. The advent of microelectronics brought about accelerometers with the charge amplifier built into the transducer (Figure 1.2). The amplifier requires a 4-20 mA constant current source to power it, and now the output from the accelerometer is voltage sensitive! In addition, the cable runs are not critical at all. The signal can run through 500 ft of coaxial cable, over railings, in heat and cold, around bends, and so on, and there is no decrease in signal strength.

Figure 1.3   Endevco charge amplifier.
(Image courtesy of Endevco, Inc.)

CASE STUDY

The U.S. Army Armament Research, Development, and Engineering Center (ARDEC) at Picatinny Arsenal, in Dover, New Jersey, was one of my customers. ARDEC used racks and racks of the Endevco Model 6634 Charge Amplifiers. My desire was to get the contract for selling the accelerometers. ARDEC would glue one accelerometer to the tip of a projectile with a mile of fine wire. Fire in the hole! Scratch one accelerometer. Next! What a contract! Disposable accelerometers—and in those days they weren't cheap.

Accelerometers have become reasonably inexpensive, with industrial grades having a wide useful frequency range from 1 to 5,000 Hz. There are accelerometers that will go to DC for very-low-frequency analysis.

Accelerometers that are stud-mounted to machines for constant monitoring require the mounting surface to be spot faced or machined flat. The facing and the stud must be perfectly orthogonal to each other. If the accelerometer is tightened down onto a cocked stud, the base will have an uneven stress applied to it. This will yield erroneously high signals. Remember, the crystal reacts to stress on its base. For this reason, all stud-mounted accelerometers should be tested with the machinery off for baseline signals, because these signals will reflect any strain at the base.

High-frequency accelerometers are responsive to 10 kHz. The accelerometer in Figure 1.4 is approximately ⅜ inch across the flats. The smaller the crystal, the smaller the mass and the higher the frequency range (Figure 1.5).

Figure 1.4   Miniature high-frequency accelerometer.
(Image courtesy of PCB Piezotronics, Inc.)

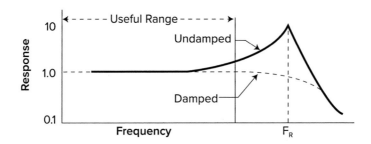

Figure 1.5   Useful range of accelerometers.
(Source: *Understanding Vibration*. Northvale, NJ: Nicolet Scientific Corp., 1980)

SEISMIC TRANSDUCERS

The predecessor to the accelerometer in the vibration field was the seismic transducer, or velocity transducer (Figures 1.6 and 1.7). The basic components consist of a magnet suspended from a spring that vibrates when the housing is mounted to a machine. The magnet moves back and forth past a coil, inducing a voltage in that coil.

Figure 1.6   IRD 544 Seismic Transducer (*top*) with a shaft stick (*bottom*).
(Images courtesy of IRD LLC.)

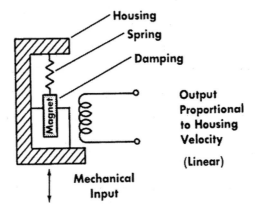

Figure 1.7   Internal construction of a seismic transducer.
(Source: *Understanding Vibration.* Northvale, NJ: Nicolet Scientific Corp., 1980)

Early steam turbine generators had transducers called *shaft riders*. These were seismic transducers mounted to the bearing housings that had a physical connection with the shaft journal through a spring-loaded nonmetallic tip that would ride the shaft, giving the technician an idea of shaft clearances in its sleeve bearing. Field technicians would attach a wooden shaft stick to the seismic transducer for taking readings directly from a rotating shaft. The strobe light would be used to make sure there wasn't a shaft key at that location. This was never a safe way to measure shaft vibration, but, yes, been there, done that.

The output of the coil is proportional to the velocity of the housing. Another benefit of velocity transducers is that the voltage is self-generating, not requiring a power-supply source. However, velocity transducers also have major drawbacks. They have nonlinear regions on both ends of their useful frequency range. They are not accurate below 10 Hz or above 2,000 Hz (Figure 1.8). They are heavily damped so that the spring-magnet system is protected from destruction by high-amplitude, low-frequency vibration. These devices are also very large and not practical for testing small machinery. Magnetic bases are also very large by necessity.

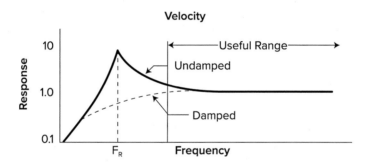

Figure 1.8   Useful frequency range of a typical seismic transducer. (Source: *Understanding Vibration*. Northvale, NJ: Nicolet Scientific Corp., 1980.)

There are applications where the customer wants a velocity output but doesn't want the frequency limitations of the standard-design seismic transducer. There are accelerometers that have a built-in signal integration circuit. With this, you have the benefit of the wide frequency range of the accelerometer and a direct velocity output. These devices are usually permanent installations with readouts going to a control room for monitoring.

TRANSDUCER MOUNTING

The internal frequency response of all transducers demands proper installation to achieve accurate results. Notice that the most widely used transducer, the accelerometer, has a natural frequency that is above its specified useful range. For example, an accelerometer that has a maximum frequency range of 5,000 Hz might have an internal natural frequency of 10,000 Hz. If the transducer is mounted with a very soft support—for example, using dental cement or a weak magnetic base—that natural frequency might drop down as low as 2,000 Hz. This would make all readings above 2,000 Hz uncalibrated (Figure 1.9).

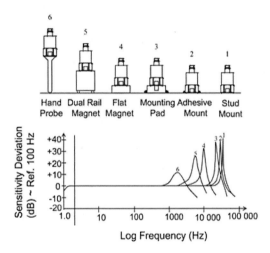

Figure 1.9   Reduction in maximum frequency range as a result of poor mounting.
(Image courtesy of PCB Piezotronics, Inc.)

DEMODULATED SPECTRA

After decades of using accelerometers for taking data, the internal natural frequency was considered to find the earliest predictions of events that impact machinery. These readings require specific considerations when mounting the accelerometer. They have different names depending on the vendor, but they are basically the same information. Rockwell/Entek's trade name for it is *Spike Energy*. Emerson/CSI calls it *Peak Vue*. A demodulated spectrum is a reading that takes advantage of the resonance of the natural frequency of the accelerometer. Natural frequencies of objects will ring or resonate when energy excites them. This is an amplified signal. For example, if a bearing has a bad outer race, as the balls run over this damaged area, they will create impacts at the ball pass frequency of the outer race. This is a predictable frequency based on the geometry of the bearing. These impacts will force the accelerometer's natural frequency into resonance. Usually, frequencies above the useful range of the accelerometer are filtered out of the reading, but they are still there.

A demodulated spectrum is the result of taking the entire time waveform of the accelerometer and first filtering out the lower-frequency energy that is predominant in the machine. The information of interest is above 5 kHz (300,000 cpm) if the accelerometer has been properly mounted (stud mounted) (Figures 1.10 and 1.11).

Once this energy has been filtered out, the remaining energy in the raw time data will be ringing energy that excites the natural frequency of the accelerometer.

Several filters are used to eliminate the lower-frequency data. If the technician or engineer suspects that the natural frequency of the accelerometer has been lowered significantly, then a filter can still be chosen to accurately identify where the resonant energy resides. A high-averaged acceleration spectrum will indicate the frequency to which the accelerometer resonance has dropped. Changing the y-scale to logarithmic will make it more obvious. It is this region that will hold the information for best results from the demodulated spectrum.

The final filtered time waveform is then rectified. This result is then put through the fast Fourier transform (FFT) algorithm, yielding a spectrum of the frequencies that are causing resonance in the accelerometer.

CASE STUDY

The data in Figure 1.12 were recorded on a large direct-drive fan powered by a motor on a variable-speed drive. The inboard fan bearing was a type that has the inner sleeve bearing liner secured by a large hex-head bolt. The bolt had worked loose, resulting in the impacting of the shaft and sleeve inside the bearing housing. Retorquing of the hold-down bolt restored the vibration to satisfactory levels. The figure shows before and after results. Importantly, other vibration parameters did *not* identify this problem.

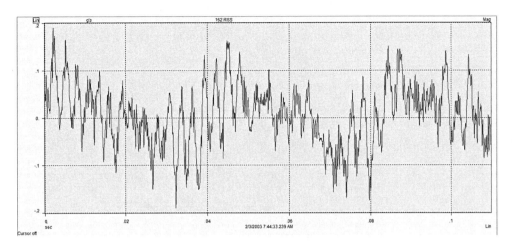

Figure 1.10　Raw acceleration-time waveform.

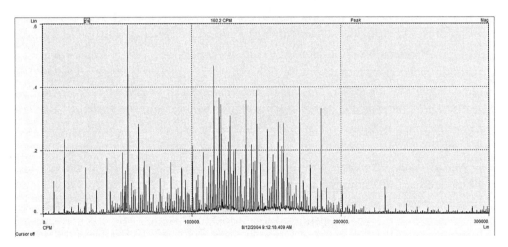

Figure 1.11　Acceleration spectrum illustrating high-frequency content.

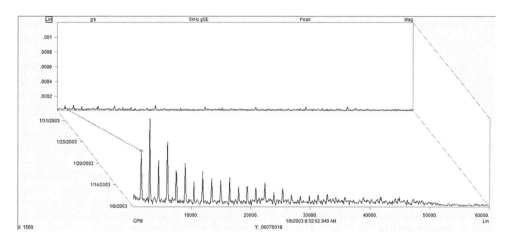

Figure 1.12 Demodulated spectrum before and after repair.

Note the amplitude units. The unit type is shown as "*g*'s," which stands for *gravitational force equivalent*, or *g-force*. This is the result of calculations being performed on the raw acceleration data. Demodulated spectra do not have normal engineering units associated with their results. For example, Emonitor Spike Energy spectra have an amplitude of gSE™. The spectra can be compared to assess a better or worse condition, but the resulting levels are too dependent on speeds, loads, stiffness of the transducer connection, and so on for there to be any value in trending levels for prediction of maintenance intervals. The most reliable method for trending is to use stud-mounted accelerometers and then to ensure that the data are recorded with the equipment operating under the same conditions.

Demodulated spectra are also highly localized, making them excellent for determining the precise location for impacting events in a machine. For example, a small motor might exhibit the same vibration energy on both ends in the velocity and acceleration data, but the demodulated spectra will differentiate the conditions. Knowing which bearing failed first could be useful for eliminating a reliability issue. However, whenever a motor is being disassembled for bearing replacement, all bearings should be replaced at the same time.

PROXIMITY PROBES

Proximity probes, or noncontact probes, provide a means of identifying where a shaft journal is running inside a sleeve bearing. The probe operates on the principle of generating an electrical field between the tip of the probe and the target shaft. As the shaft moves within this field, the interaction within the electromagnetic field is converted by the proximeter to an alternating-current (AC) voltage that is proportional to the displacement of the target. Most probes have a sensitivity of 200 mV per mil of displacement, peak to peak, and a linear useful range of 0.070 inch. The DC voltage component from the proximeter is first used to set the probe gap to the target and can yield dynamic information about the position of the shaft in the bearing. They can be utilized to watch axial movement as well to ensure the thrust bearings in a sleeve bearing machine are operating within specifications. The probes are placed in staggered pairs for the thrust-bearing reading. If one probe comes in contact with the thrust disc because of a failing thrust bearing, the monitoring system can still use the second probe to safely shut the machine down (Figure 1.13).

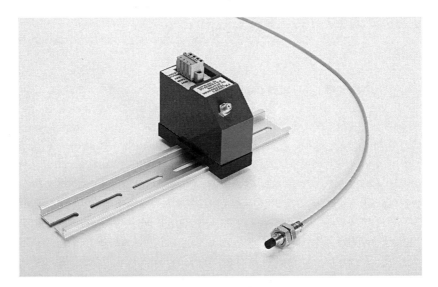

Figure 1.13   Proximity probe and proximity driver.
(Image courtesy of SEC of America, a US subsidiary of Shinkawa Electric Co.)

Warning: Proximeters can have connectors from which technicians can take direct readings. When the output is used for emergency shutdown of the machinery, disconnecting the output cable will send a large DC spike to the readout, and a trip will occur. Always place monitoring systems into a temporary "monitor-only" state before disconnecting the proximeter output cable.

CASE STUDY

I was sent to a process facility to read the air compressors and ancilliary equipment on a monthly basis. The route was fairly simple, requiring approximately 1 hour to complete. There were two large air compressors that required readings from the proximeters. The site manager was notified so that the automatic shutdown in the software would be disabled. On one particular day, the control room was very crowded due to coordination of the installation of a large piece of equipment. The message to disable the shutdown was lost in the confusion. I went to the first compressor, opened the proximeter cover, and disconnected the BNC (Bayonet Neill–Concelman, named for the inventors of the cable locking device, Paul Neill and Carl Concelman) at the proximeter. The machine instantly shut down. The site required 10 truckloads of product delivered that day to meet demand. It took 10 hours to get the site process stabilized and operating properly.

These probes require calibration for different shaft materials. They are calibrated to their length of pickup cable and require highly polished and burnished target areas on the shaft journals, as well as degaussing to remove magnetic fields. Engineering design requires them to confirm bearing stability. Consider that electric motors have a rotor that is more than 33% of the total weight. As a result, vibration on the bearing caps is representative of what is important to trend, and therefore, accelerometers are more than satisfactory for trending. Proximity probe output is limited by the probe's dynamic range to a useful frequency range of the seventh harmonic of 1× running speed. For this reason, accelerometers are also required for gearboxes, compressors with a high number of vanes, and so on.

Machines with a rotor-weight-to-machine-weight ratio that is high—for example, an air compressor with small, high-speed turbo wheels—must use proximity probes to monitor bearings and gear clearances. A Centac compressor, for example, has compressor wheel pinion shafts that weigh 10 pounds, whereas the machine total weight is 7 tons. The wheels are geared up to as high as 80,000 rpm (Figures 1.14 and 1.15).

Proximity probes are essential to catch the onset of bearing wear. Missing this event would result in a catastrophic failure of the compressor, costing thousands of dollars more than a bearing replacement/overhaul.

Figure 1.14   High-speed compressor wheels on a Centac compressor.
(Image courtesy of Ingersoll Rand.)

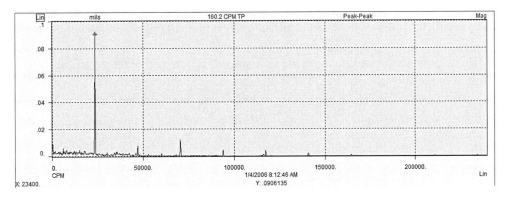

Figure 1.15   Spectrum result from proximity probe: 1× running speed is 23,400 rpm.

The plot in Figure 1.15 was taken with a proximity probe looking at a compressor wheel shaft running at 23,400 rpm. This is typical of the types of machines where the use of proximity probes is essential. Notice that the vibration level of 1× running speed is <0.1 mil, peak to peak, of displacement. Notice that the compressor also has high tooth gearing, which require 10-kHz accelerometers that are stud mounted to see the gear mesh frequencies.

The example in Figure 1.16 illustrates why time-domain signals are important for diagnosing machinery issues. The peak values in this plot are approximately 40 *g*'s, peak to peak. The next plot is the FFT of this time waveform. Remember, the FFT process considers the energy under the curve. These peaks are very narrow, not like a perfect sine wave.

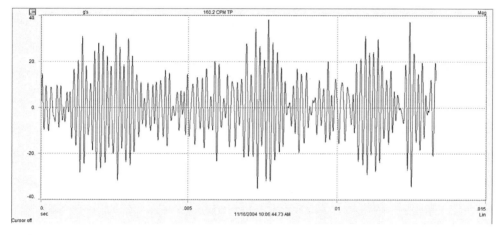

Figure 1.16   Time waveform of a third-stage pinon using a 10-kHz accelerometer.

The peak values in the spectrum in Figure 1.17 are just above 5 g's. This is the result of the root-mean-square (RMS) calculation on narrow spikes in the time waveform. The 20-g peak of the narrow time spike has only 5 g's of energy under the curve. If the data collected had not included the time waveforms, those high-impacting levels would not be known.

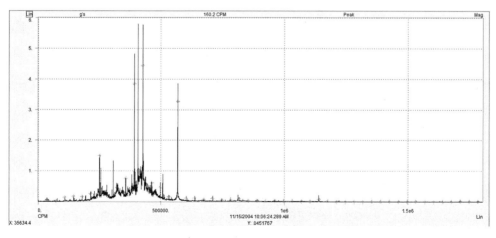

Figure 1.17   Acceleration spectrum of third-stage pinion.

The cursor in Figure 1.18 is positioned at 3,600 cpm, running speed of the motor, and the bull gear of the compressor. The *x*-axis is in orders of running speed, "1X" marks running speed. The speeds of the pinions are in the 20–40 order range on the plot. The energy above 70 orders is affected by the accelerometer resonance dropping into the frequency range of the plot because of magnetic attachment. These units did not have stud-mounted accelerometers. The reader can see how low the levels of the 1× running speed of the motor and pinons are to understand why casing readings are not acceptable for reliability trending of high-speed compressors.

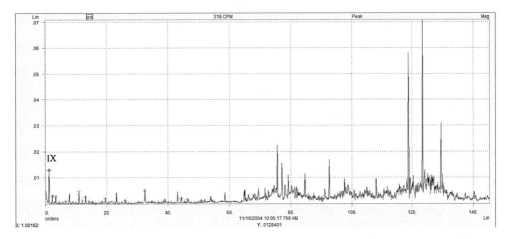

Figure 1.18   Velocity spectrum of third-stage pinion.

ORBIT ANALYSIS FROM PROXIMITY PROBES

The top plot illustrates a satisfactory orbit (Figure 1.19). The bottom plot identifies a misalignment in a machine by viewing the orbit created by viewing two proximity probes that are mounted at orthogonal angles from each other. The two probe signals are viewed in an oscilloscope using x-y coordinates. Misalignment is usually diagnosed by a 2× increase in running speed vibration. The bottom figure has a 2× component. Two single-frequency sine waves of equal amplitude would generate a perfectly circular orbit. The magnitudes are not equal in the orbit on the top, generating an ellipse. These magnitudes are normal for field-installed equipment because the horizontal direction is normally less rigid than the vertical direction.

The output of the two proximity probes in Figure 1.20 also can give a magnitude/ phase relationship of the shaft centerline during a run-up or a coast-down of rotating equipment. These diagrams are helpful in identifying natural frequencies and their phase relationships.

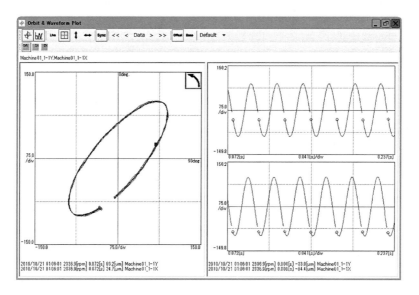

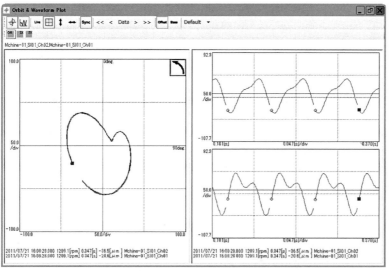

Figure 1.19   Acceptable orbit (*top*) versus misalignment (*bottom*) utilizing proximity probe orbit.

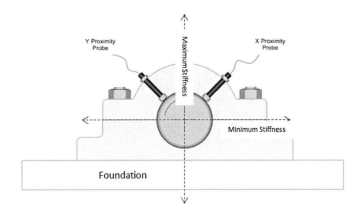

Figure 1.20 Proper installation of two proximity probes for orbit diagnosis.
(Image courtesy of SEC of America, a US subsidiary of Shinkawa Electric Co.)

The Bode plot (*right*) and the polar diagram (*left*) in Figure 1.21 are the result of saving the 1× running-speed component and its associated phase reading, as well as a speed reference and plotting them over the run-up or coast-down of a machine. These examples clearly indicate that a natural frequency (resonance condition) is encountered near 1,538 rpm.

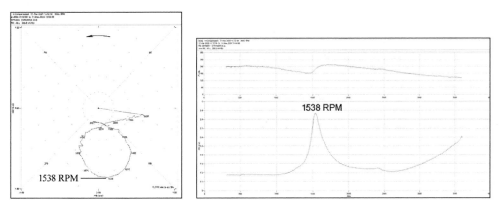

Figure 1.21 Bode plot with associated polar diagram.
(Image courtesy of Jeff Kenney, IPS.)

Fortunately for the technicians of today, modern FFT analyzers are technically advanced beyond the limitations of their predecessors. Technical specifications of the 1970s included analog-to-digital conversion, which limits the dynamic range of the captured data. Imagine having to divide a 5,000-Hz maximum frequency range into only 200 lines! That's a frequency resolution of only 1,500 cpm:

$$5,000 \text{ Hz } F_{max}/200 \text{ lines of resolution} = 25 \text{ Hz/bin}$$

$$25 \text{ Hz/bin} \times 60 = 1,500 \text{ cpm}$$

This means that moving the cursor one bin will change the readout from 1,500 to 3,000 cpm. The analyst will have all the vibration between those two frequencies in one bin. That is almost as bad as doing an analysis with just an overall reading. By necessity, several zoom ranges were required to analyze the entire frequency range of the machine. Fortunately, 400-line resolution was the standard by the mid-1970s. Today's analyzers can take 16,000 lines of resolution. Long averaging times are the tradeoff.

One specification that has gone backward is real time rate. The Nicolet 100A Mini Analyzer had a real-time rate of 10,000 Hz, meaning that any change in the vibration occurring at that frequency or less would be captured and viewable without any gaps in the data. This was exceptional for monitoring captured data during a run-up or coast-down. Modern FFTs capture data at rates of 5,000 Hz, but the screen update rates are much slower. For the most part, everything is post-capture diagnosis.

The old argument was that without the fast updates, you might miss transient spikes. Unfortunately, transient data are a completely different issue and should be captured and plotted using triggering on the input signal. Transient data are not useful for normal trending. Triggering should have a delay as part of the setup. The delay will put the transient data in the middle of the data screen instead of at time 0, when the trigger occurred.

Figure 1.22 illustrates a 40% delay on an impact capture. A 10% delay will work well, leaving 90% of the window to view the ring-down. Notice the background vibration at the beginning of the waveform, captured while the analyzer "waited" for the impact.

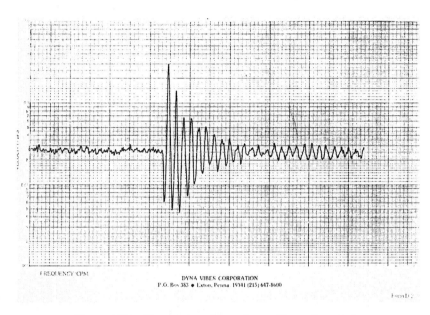

Figure 1.22   Time-delayed waveform of impact analysis.

SOME EXAMPLES OF KNOWING THE FUNDAMENTALS

Not all field service calls require a trunk load of test equipment. These "easier" calls can be the most frustrating for customers, so the field service technician should use the utmost of tact and politics to pass along the news.

CASE STUDY

A large utility customer called for a technician to travel across the country to troubleshoot a high vibration on its main steam turbine. Vibration levels were very high on the generator end of the machine. The total travel time to the job was 9 hours (two plane rides and a ferry ride). The technician walked up to the proximity-probe outputs in the control room, looked at the reading, which was over 10, and threw the switch on the front of the panel. The reading dropped to 1. Someone at the utility had checked the DC gap on that proximity probe and forgot to switch it back to vibration. The "10" reading was 10 volts. This is typical for centering a proximity probe at the center of it's 0.070" linear range. There was no vibration problem! The plant engineer was not happy. Neither was the technician. These are the kinds of mistakes everyone in the industry learns from. With experience, questions could have been asked about the setup of the readouts over the phone, and much time, trouble, and money could have been saved. Return trip? Twenty-one hours with delays.

CASE STUDY

An original equipment manufacturer (OEM) had received a new motor and was solo testing it on the shop floor. The OEM engineer called to say that the vibration did not meet the company's acceptance criteria. Questions were asked: "On isolation pads?"; "Yes"; "Coupling installed?"; "No"; "Half key installed?"; "No." The OEM had nothing on the motor shaft, just a bare keyway. They were instructed to fill the keyway with two half-sized pieces of key stock and secure them with hose clamps and then cover them with glass tape. The OEM engineer said, "Oh, come on. The vibration is 0.3 inch/s velocity. That's going to make a difference?" The engineer followed through with the request and then called back the next day. "Well, we learned something today. The vibration levels dropped to 0.03 inch/s velocity!"

Caution: Key stock used for testing a motor without its coupling must be secured to the shaft with two steel hose clamps and then covered with duct tape or glass tape. Place the hose clamp screw blocks 180 degrees apart to balance them. End the tape wrap close to the starting point to minimize imbalance.

CASE STUDY

A power plant was getting ready to test a 2,500-hp, 3,600 rpm motor without its coupling. I was informed that the proper key stock had been put in place on the bare shaft. Plant personnel were removing the locks and racking the motor breaker back in for a test run. I was making sure the engineer could fit between the machines to get a vibration reading on the motor inboard bearing. This is when I noticed that there were no hose clamps on the key, just duct tape. The motor startup sequence was halted immediately. The key stock was 1⅛ inches square and >7 inches long. This is a mass of >2.5 lb. Imagine someone getting hit by a 2.5-lb mass traveling at >60 mi/h. The plant engineer then explained that someone from the "other" motor repair shop had put the tape on the key stock and was taking readings the previous day. Obviously, that person had his guardian angel looking over his shoulder. Safety should always be first in the minds of everyone.

What vibration frequency can we expect to get from an unbalanced situation? The vibration will be at the same frequency as the running speed of the rotor. The narrow band signature in Figure 1.23 has a large percentage of the vibration occurring at 1× operating speed of a fan that requires balancing.

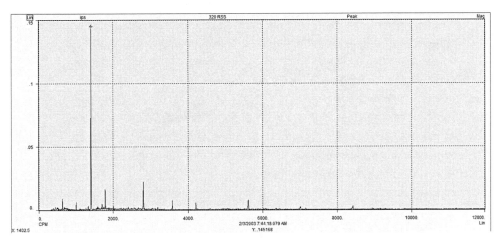

Figure 1.23   Unbalanced vibration at a running speed of 1,400 rpm.
(Image courtesy of Ron Brook, IPS, 2003.)

C A S E   S T U D Y

A printing company was experiencing high vibration levels on the entire printing line. The vibration was so severe that the machine could not hold register, and the colors smeared. The frequency of the vibration was identified, and the technicians then went about looking for any rotor turning at that speed. They ended up on the final section of the machine where the pages were cut for folding. There was a large cutter bar bolted into the roll. This was seen using a strobe light. The frequency of the strobe light was changed slightly to slowly roll the rotor to view the entire circumference of the roll. At 180° from the bolted cutter bar was a deep, open slot. The maintenance manager was consulted, and he started to explain that they were cutting at one-half the rate on this particular run, so one of the cutting bars was removed. The realization hit him that a 10-lb cutting bar had been removed, but the 10-lb blank had not been bolted into the rotor in its place. The rotor had 10 lb of imbalance!

CASE STUDY

A funny side note to that job: The press was a four-color machine. A vibration technician was taking readings on a sensor that rode on the paper at the first color station. He slipped and his vibration probe tore the paper, resulting in the paper having to be refed through the entire machine. The customer congratulated him on hitting a home run. Because there were four colors, if they only had to feed the back section, that was a single; two sections, a double; and so on.

Normally, machinery vibration is not purely sinusoidal, so don't expect one single peak in the FFT. There can be times when the unbalance is so severe, and the system has little or no damping that the running-speed peak (1×) will be a predominant peak on the plot.

As seen above, the 1× running speed is predominant, but there are still other peaks that are associated with other mechanical and electrical issues in the machine. There are also harmonics of running speed in the preceding spectrum.

These harmonics will normally taper off by the seventh harmonic. A machine that has excessive hammering (i.e., loose bearing fits, cracked rotors, etc.) can generate harmonics as high as 25–40× running speed. This brings up an important point about reducing vibration. Alleviate the primary cause of the vibration, and many times the remaining harmonic vibration will be alleviated as well, driving down the overall vibration level.

There are other examples where removing one problem makes no difference to the vibration level of another source. Sometimes eliminating one fault will unmask another problem that may have been the main issue all along.

REVIEW QUESTIONS

1. Give three reasons why the ICP accelerometer is the chosen transducer for vibration analysis.

2. Which industrial transducer does not require a power supply (self-generated signal)?

3. True or False: The resonant natural frequency of a seismic transducer is in its lower useful range.

4. What are the negative aspects of using a seismic transducer?

5. What can be used for a velocity output without the negative aspects of a seismic transducer?

6. What type of signal is the output of a non-ICP accelerometer?

7. Why is an ICP accelerometer better than a non-ICP accelerometer?

8. What is the most accepted input power required for an ICP accelerometer?

9. True or False: The larger the accelerometer crystal, the lower the maximum frequency range.

10. What are the pros and cons of measuring and trending demodulated spectra?

11. What does the unit G_{SE}™ represent on a demodulated spectrum? Is it relevant to other vibration parameters?

12. Where is a proximity probe use essential to identifying a problem?

13. What does a field technician need to be aware of when disconnecting the BNC connector on a proximeter of a loaded, running machine?

14. A proximity probe has a useful frequency range related to what operating parameter?

15. What is the optimal distance a proximity probe should be mounted from the shaft of interest?

16. What does a technician measure to set a proximity probe into a machine for the most useful range?

17. A proximity probe, with a sensitivity of 200 mV/mil, yields a reading of 1 volt. What is the shaft displacement at this probe?

18. What parameter in a vibration spectrum is most important for separating individual fault frequencies?

19. What are two key measurements a field technician can see at a proximity probe readout?

20. True or False: The rotor vibration of an electric motor cannot be read without proximity probes.

21. True or False: Accelerometers must be used instead of proximity probes to see gear mesh frequencies.

22. Why is it important to know which parameter you are reading on a proximity probe panel meter?

23. What will happen to the vibration level of a rotating shaft if there is a mass change at one radial location?

24. What type of mounting for an accelerometer yields the highest frequency range transmissibility?

25. What conditions must be met for stud mounting an accelerometer?

CHAPTER 2

Waveforms, Spectral Domains, Orbits, and More

WAVEFORM NOMENCLATURE

Imagine the rise and fall of a ripple on a lake. That ripple has peaks and valleys. There is also a frequency to how close together or far apart the peaks and valleys are. We can think of the raw vibration energy from a machine as having this same ripple, or waveform.

A single frequency is represented in Figure 2.1 in its time waveform. The waveform displays the vibration magnitude (y) as a function of time (x). Vibration definitions use peak-to-peak values for displacement readings because this captures the total motion of the target. Velocity is measured in peak units because the maximum velocity is the same above and below the zero line, with just a change in direction. Acceleration is measured in peak-to-peak units. European definitions will use root-mean-square (RMS) for acceleration. This is the area under the curve. For a pure sine wave, the area under the curve is equal to 0.707 times the peak value.

- Peak to peak: Maximum negative peak to maximum positive peak
- Peak: Zero to maximum positive or negative peak
- Average: 0.637 times peak (average volts = 2 × peak/π)
- RMS: 0.707 × peak

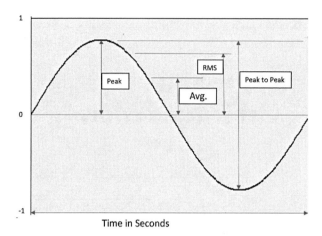

Figure 2.1 Time waveform definitions.

COMPARING THE THREE MAIN VIBRATION PARAMETERS

The nomograph in Figure 2.2 illustrates the relationship between the three most common vibration measurement parameters. This graph makes it obvious that measuring velocity yields the most useful information concerning the severity of the vibration because it is evenly weighted across the entire frequency range of interest.

A common mistake of field technicians is to put too much emphasis on high-frequency vibration information by viewing the spectrum in acceleration. The values of 20 or 30 g's at a gear mesh frequency sound alarming, but the displacement is usually less than 0.001 inches!

Before the proliferation of heldhand calculators and smart phones, vibration parameter calculators were handy tools that were great giveaway items at trade shows and sales presentations (Figure 2.3). The user would slide the inner sleeve until the speed of the machine was under the arrow indicator. Then the user could compare the vibration reading with any of the other parameters, as well as English or metric units.

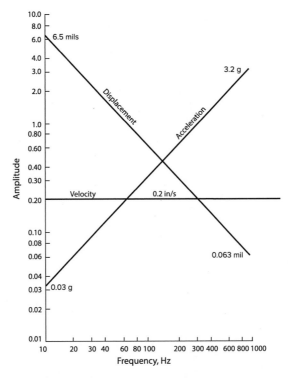

Figure 2.2 Vibration measurements nomograph.
(Source: Nicolet Scientific, Understanding Vibration, 1980.)

The back of the calculator had decibel to linear unit and mils to micrometers converters and Centigrade to Fahrenheit conversion charts. All the formulas for converting the three vibration parameters were there as well.

Figure 2.4 illustrates why accelerometers are so widely used. The shaded arrows on the high and low ends of the accelerometer range indicate that there are special accelerometers to cover those ranges, as low as direct current (DC) and for higher frequencies (up to 10 kHz).

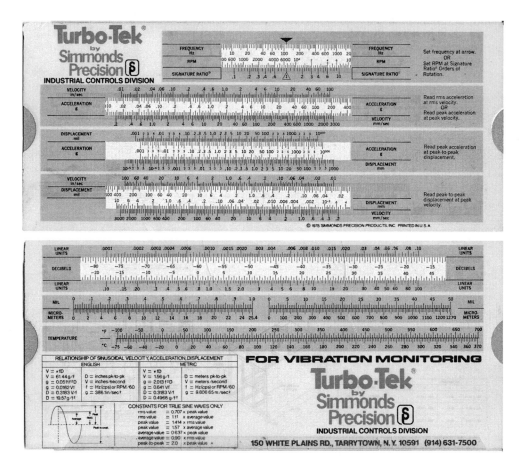

Figure 2.3 Vibration parameter calculator.

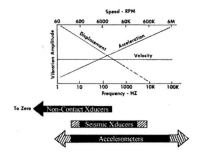

Figure 2.4 Useful range of transducers.

(Source: Nicolet Scientific, Understanding Vibration, 1980.)

Let's discuss the three predominantly used vibration parameters. Here we will use a piston in an internal combustion engine to illustrate time waveforms and how displacement, velocity, and acceleration relate to each other. Figure 2.5 plots the movement of the piston as it moves from its midpoint to top dead center, back through the midpoint, down to bottom dead center, and finally returning to the midpoint. The displacement of the piston is referred to in peak-to-peak terms—that is, from top dead center to bottom dead center.

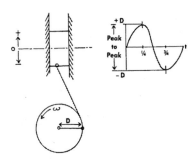

Figure 2.5 Plotting displacement over time.
(Source: Nicolet Scientific, Understanding Vibration, 1980.)

The velocity of the piston is maximum as the piston passes through the midpoint in both directions (Figure 2.6). The velocity at top and bottom dead center is zero. Plotting the two graphs over each other yields the plot on the right. The two curves are 90 degrees out of phase with each other, velocity leading displacement by 90 degrees.

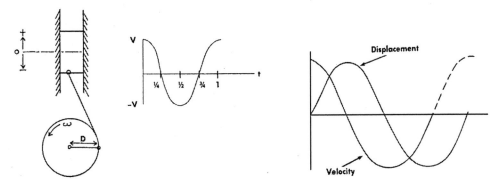

Figure 2.6 Velocity of a piston.
(Source: Nicolet Scientific, Understanding Vibration, 1980.)

The acceleration of the piston at the midpoint is zero (Figure 2.7). This is because there is no change in velocity at that instant in time. The maximum acceleration and deceleration occur at top and bottom dead centers as the piston comes to rest and then reverses direction.

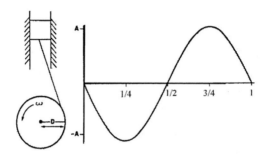

Figure 2.7   Acceleration of a piston.
(Source: Nicolet Scientific, Understanding Vibration, 1980.)

Therefore, we normally refer to velocity as leading displacement by 90 degrees, and acceleration is 180 degrees out of phase with displacement (Figure 2.8).

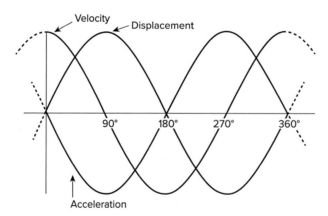

Figure 2.8   Displacement, velocity, and acceleration time waveform overlays.
(Source: Nicolet Scientific, Understanding Vibration, 1980.)

A word about spectral analysis: the fast Fourier transform (FFT) takes the input time waveform, separates out all the frequency content in that time waveform, and then displays each separate frequency component as a single peak. The FFT does not get generated from left to right in time, like the time waveform. The x-axis of a spectrum is frequency, not time.

The three graphs in Figure 2.9 illustrate how the three vibration parameters differ from the same measurement. Acceleration is more sensitive to higher-frequency events in the machinery. Velocity also contains that higher-frequency content, as well as the lower-frequency content. Displacement has none of the higher-frequency information but enhances low-frequency peaks.

The examples in Figure 2.9 have been filtered on the low-frequency end to eliminate what all technicians know as the *ski slope*. This is a large curve at the low end of the spectrum, and it is the result of digital integration within the hardware. The integration of a peak value at 0 is infinity. The low-end peak can be so great that the rest of the scaled data are impossible to interrogate.

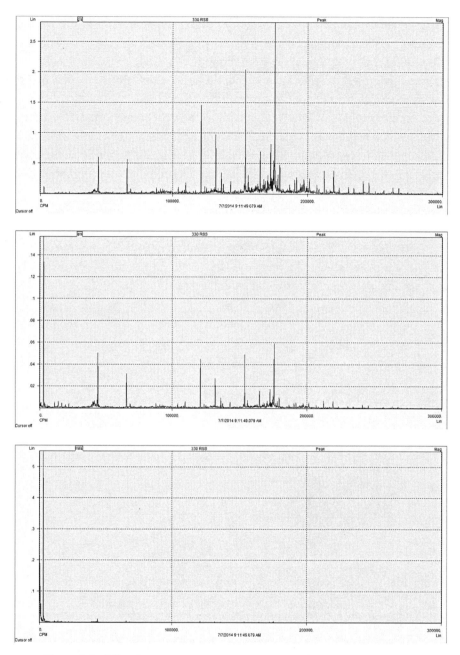

Figure 2.9   Vibration measurement in acceleration (*top*), velocity (*center*),
and displacement (*bottom*).

PITFALLS TO AVOID WITH FFT SETTINGS

Readers can see the values for the 900-rpm peak on the bottom of the plot in Figure 2.10. Notice the amount of error between the two spectra (Figures 2.10 and 2.11). A 0.02 peak is erroneously listed as 0.1. That's a 500% error!

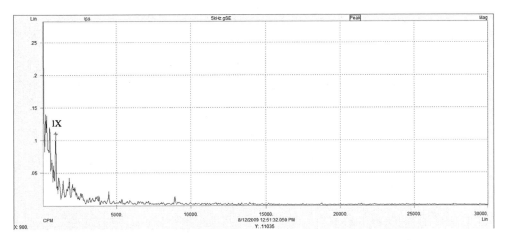

Figure 2.10   Velocity spectrum from integrated acceleration spectrum
without low-frequency filtering.

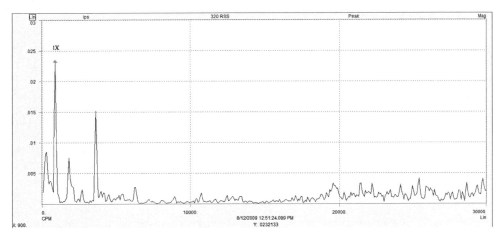

Figure 2.11   Velocity spectrum with 320 cpm filtering of the same accelerometer input signal.
Note: Text "1X" marks the running speed of the motor/fan at 900 rpm
in Figures 2.11 through 2.14. This would be a peak of interest to a technician.

Double integration of the accelerometer data doubles the problem of the slope. An uninformed technician would see the peak at 70 mils and report this as a major problem with the machine (Figure 2.12).

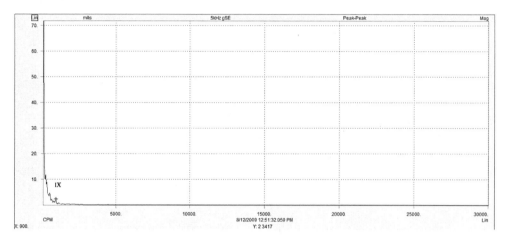

Figure 2.12 Displacement spectrum as the result of double integration of the accelerometer spectrum.

The filtering of the acceleration input for the velocity spectrum yields a usable integrated displacement spectrum. The 70-mils reading has been removed, and all peaks are now below 1.0 peak to peak.

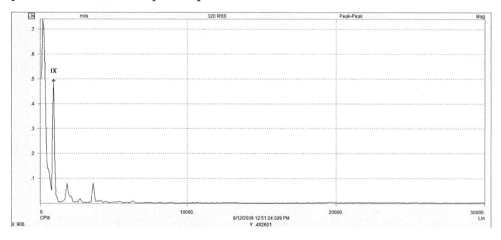

Figure 2.13 Displacement spectrum as a result of single integration of the filtered velocity spectrum.

The details in selecting the parameters for capture of the vibration data are obvious. Data captured as raw acceleration without low-end filtering yields little useful data. The velocity spectrum was plotted for example purposes only. The technician should record several spectra to maximize the effectiveness of the data. The lower-frequency vibration data should be captured with a typical maximum frequency of 12,000 cpm and 800 lines of resolution. The velocity spectrum in Figure 2.11, as a result of expanding the frequency view from 60,000 to 30,000 cpm, has only 400 lines of resolution.

Notice that with more lines of resolution, the spectrum frequencies are better defined (Figure 2.14). This is discussed again when we identify mechanical versus electrical vibration data.

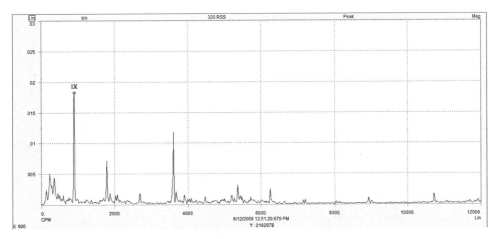

Figure 2.14   Velocity spectrum captured with 12,000 cpm
maximum frequency and 800 lines of resolution.

Note: An adequate filter for velocity measurements is 5.5 Hz, or 330 cpm. Readers will note the filter at the top center of the plot is 320 RSS. Emonitor is designed to minimize the time taking data on routes by using these filters. The 320 RSS filter will not only provide for slope-less velocity data from the accelerometer but also will automatically calculate the overall velocity level from the spectrum. This negates having to build "overall" collections into the database (Figure 2.15).

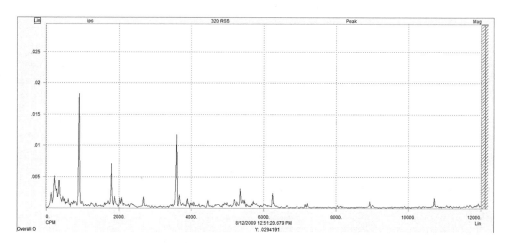

Figure 2.15   Velocity spectrum with overall calculated band (*far right*).

The overall has been calculated by operating on the captured data and is displayed at the bottom of the plot. Overall level is 0.0294191 inch/second. The software also corrects each peak value for the windowing applied to the data. A Hanning window will reduce the peak values by approximately 50%. Therefore, each peak value is multiplied by 2 before displaying. The overall value represents all the energy under the curve of the FFT (RMS calculation).

BEAT FREQUENCIES

Technicians may encounter fluctuating or pulsing vibration levels. These can be non-periodic step changes or have a periodic beat to them. Periodic beats are referred to as *beat frequencies*. They occur when there are two closely spaced vibration frequencies at the same measurement point. Two machines, side by side, may have the same nameplate speeds on the motors, but there will still be a slight difference. The time waveforms in Figure 2.16 are an example of what a beat frequency looks like.

Notice that the two original signals are of equal amplitude. The only difference is the frequency. The vibration magnitudes add when they are in phase, and they subtract when they are out of phase. The beat frequency is the difference between the two frequencies. If one motor is operating at 3,585 rpm and the other motor is operating at 3,570 rpm, the beat frequency will be 15 cpm. The amplitude fluctuation the analyst measures, hears, and feels will occur every 4 seconds. Beat phenomena are common

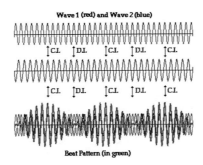

Figure 2.16   Beat frequency.
(Source: http://www.physicsclassroom.com/Class/sound/u11l3a6.gif.)

in induction motors because the rotor rotation is slower than the field. A four-pole motor with a misalignment problem will generate vibrations at 7,170 cpm (due to the misalignment) and 7,200 cpm (two times line frequency). The beat frequency will be 30 cpm.

To view a beat-frequency magnitude live using an FFT, use less resolution. With high resolution, the user will see the two individual peaks without any magnitude fluctuations. With less resolution, the two peak-phase interactions can be viewed on the FFT as a modulating amplitude. This is due to the phase change residing in one or two bins of the FFT. When the two peaks are separated by a sufficient number of bins, they each have their own phases and magnitudes.

C A S E   S T U D Y

What kinds of problems can beat frequencies cause? Consider an industrial complex with 20 machines all running at the same speed, all feeding into the same process. Figure 2.17 shows a waterfall plot taken on one motor at one location over a 50-minute interval. What should the technician put down on their report for the vibration level? The magnitude is changing from 0.5 inch/second to less than 0.2 inch/second. This is all due to the motor's 1× vibrations moving in and out of phase with each other. Because there are more than two motors, the beat frequency is complex.

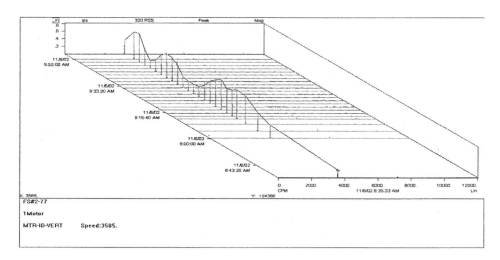

Figure 2.17   Beat frequency of motor/compressor beating
with 19 other motor/compressor units.

Synchronous time averaging can be used to balance and/or analyze a machine that is beating with other machines. This type of averaging requires a tachometer signal input into the data collector. The analyzer takes each average as the tachometer triggers on the machine under test. This results in vibrations that are *not* in phase with the machine under test to be subtracted out. This process will take many more averages than normal testing.

LABELING MEASUREMENT LOCATIONS ON MACHINERY

CASE STUDY

Working for a motor manufacturer and repair facility, you learn things. Ask any technician taking vibration data which end of the motor they would call the back end. The technician will point to the end of the motor opposite the shaft end of the motor. And if the technician called the manufacturer and ordered the back-end motor bearing bracket, they would get the wrong part! The electric motor industry has always referred to the opposite drive end of the motor as the front end. This example illustrates how important it is that the labeling nomenclature used for any vibration program is self-explanatory.

There are other examples. A visit to a refinery brought me a request to look at vibration data recorded by the company's technician on a given motor. The readings were labeled "1A," "1B," "1C," "2A," "2B," and so on. There was no way to identify where the data were recorded, and the vibration technician was off for the week. Also, no one else had been educated on what the technician's labeling protocol was.

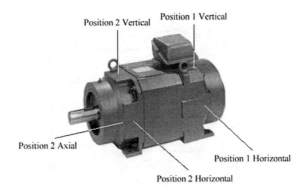

Figure 2.18   Horizontal machine measurement position labels.
(Image courtesy of Reliance Electric.)

A horizontal machine should have the labeling start at the point furthest from the driven equipment. A single-ended motor has two bearings. Position 1 in this case would be motor outboard, referring to being opposite the output shaft. The horizontal axis refers to a reading that is horizontal with the motor support feet. The vertical axis is perpendicular to the horizontal axis. The axial axis looks at vibration in line with the shaft. Examples for labeling would be "MOH" (*motor outboard horizontal*), "MOV" (*motor outboard vertical*), "MOA" (*motor outboard axial*), "MIH" (*motor inboard horizontal*), "MIV" (*motor inboard vertical*), and "MIA" (*motor inboard axial*).

Vertical machinery provides a slightly more complicated means of labeling (Figure 2.19). There are four readings that are horizontal to the ground, and the one vertical reading is looking down the shaft, or axially. The two radial readings taken at each bearing can be labeled "east-west" and "north-south," but I prefer "in-line" and "perpendicular." If the motor is mounted on the top of a pump head, the two directions can refer to the visible piping. This is very handy because most vertical motor/pump assemblies have very different spectra from these two radial locations.

The reason is the additional stiffness added by the horizontal discharge pipe. Another reference point for in-line perpendicular labeling can be the motor terminal box when the piping isn't visible.

- Motor top in-line
- Motor top perpendicular
- Motor top axial
- Motor bottom in-line
- Motor bottom perpendicular

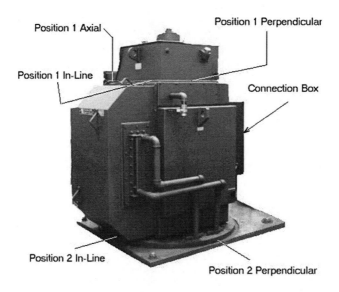

Figure 2.19   Vertical machine measurement position labels.
(Image courtesy of Reliance Electric.)

CASE STUDY

Labeling is probably thought of as trivial by some design engineers, but there was one customer site that made the haphazard labeling appear intentional. There were two machines at each site. The site name would be followed by the two machine names. The two machines at eight different sites were labeled as "1" and "2," "North" and "South," "A" and "B," and so on.

Another site had dual pumping systems labeled "East" and "West," but then there were east and west sections of the plant, so there was a "#1 west west," "#1 west east," and so on. This labeling made data collection a real chore.

Labeling a motor-driven pump requires 10 labels. There is usually only one axial reading per machine. This labeling system is concise, and anyone can figure out where each reading was recorded (Figure 2.20).

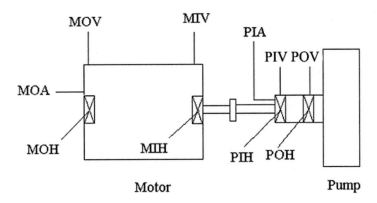

Figure 2.20   Labeling a motor-driven pump.

Phase measurements are very important to understanding how a machine is vibrating and for dynamic balancing. Figure 2.21 illustrates a handheld data collector with two inputs. One is for the vibration transducer, or accelerometer, and the other is the tachometer input. Many types of tachometers exist, but the most common one in use today is an infrared sensor that triggers on a signal returned by a piece of reflective tape placed on the shaft.

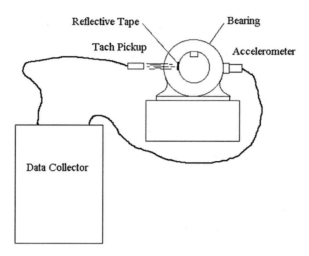

Figure 2.21   Measuring phase on rotating machinery.

Balancing requires that we know where the shaft position is when the maximum vibration is captured by the accelerometer. The trigger tells the analyzer when the shaft passes that reference point. The analyzer can then calculate how many degrees the shaft has rotated through since the maximum vibration was sensed by the accelerometer.

The two vibration signals in Figure 2.22 are 90 degrees out of phase with each other. In all other parameters, they are equal.

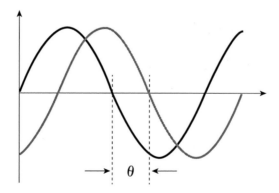

Figure 2.22   Two identical frequencies and magnitudes, 90 degrees out of phase.
(Source: Wikipedia.)

Figure 2.23 illustrates the difference between a motor with the ends in phase and the ends out of phase. This kind of information is critical when attempting to correct vibration problems. If this situation were due to imbalance, the top motor would require a static balance (i.e., the weights are placed at the same angular location on each end), whereas the bottom motor would require a coupled balance correction (where the correction weights on each end are placed 180 degrees apart from each other).

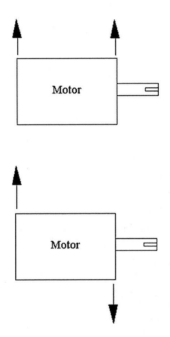

Figure 2.23   Ends of two motors, in phase and out of phase.

For practical purposes, phase relationships are referred to as *in phase* if the difference between the two is 90 degrees or less and *out of phase* if the difference between the two is 90–180 degrees.

Time-domain data from a real-world machine are never easy to read. If the waveform were a pure sine wave, the technician could count the peaks and multiply them by the reciprocal of the time base to find the frequency. This is impossible with a complex waveform such as that in Figure 2.24.

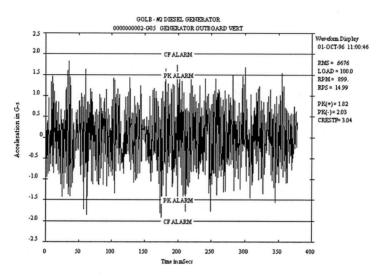

Figure 2.24   Complex time waveform.

SHORT HISTORY OF THE FFT

Joseph Fourier, the famous French mathematician, came up with the formulas to calculate the frequency components from the time domain in the early 1800s. However, his math required that you take all time points from negative infinity to positive infinity. Not likely with today's time constraints on getting jobs completed.

Enter Mr. Cooley and Mr. Tukey. These two gentlemen came up with the fast Fourier transform (FFT) algorithm while working at IBM research laboratories in 1965. In that year, Princeton University published their historic paper on computation of the FFT. This development allowed a 1,024-point FFT transform (512 lines) to be processed in one one-hundredth of the time previously required. It required 50% less memory to complete the computation ("The 'real' history of real-time spectrum analyzers: A 50-year trip down memory lane," Joe Deery, AIM & C, Newfoundland, New Jersey).

This landmark change in frequency analysis laid the groundwork for what is now the handheld data analyzer market. Before the FFT, technicians used tunable filter analyzers attached to strip-chart recorders (Figures 2.25 and 2.26). The analyst would flip a switch, and the filter would automatically sweep over the entire frequency range. The pen would trace the magnitude encountered as each frequency was centered in the filter. One drawback was the time it took for one trace to complete. But the major drawback was missing data if a frequency event happened while the filter was somewhere else.

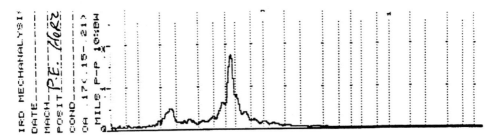

Figure 2.25 Percentage bandwidth vibration analysis.

Figure 2.26 Vic Ryan using a tunable-filter analyzer and printer.
(Image courtesy of IRD LLC.)

The process was tedious, and percentage bandwidth filters have a logarithmic frequency scale. This means the lower frequencies are spaced apart clearly, but the higher-frequency content is compacted without a clear definition of frequency peaks.

Thanks to a forward-thinking engineer by the name of Robert Leon, his field service crew at Vibration Specialty Corporation in Philadelphia was among the first technicians to take narrow-band analysis to the field to tackle vibration problems. The

Nicolet UA500 was not a digital FFT. It was an analog time-compression analyzer (Figure 2.27). There were two potentiometers that the technician would have to tweak periodically while watching an internally generated calibration signal on the oscilloscope. This was to calibrate the frequency (*x*) and magnitude (*y*) axes while watching the oscilloscope screen.

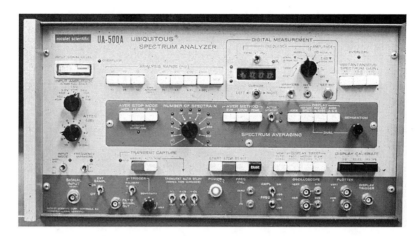

Figure 2.27 Nicolet UA500 time-compression real-time analyzer. (Image courtesy of Vibration Specialty Corporation.)

Both of these analyzers gave a huge boost to performing diagnostics in the field. The advent of the digital analyzer eliminated the need for calibration changes in the field. The configuration in Figure 2.28 included an analog inverter to measure acceleration, velocity, or displacement. The hardware also included a tape storage unit for spectra and an *xy*-plotter for hardcopy. The plotter started after the spectrum had been averaged and saved. *Field portable* was the key terminology here. It had previously unknown power but was slightly bulky. The Nicolet analyzers also had the ability to zoom the spectral data. This is not the same as expanding a fixed-resolution spectrum. The technician could take the 400 lines of resolution and apply them to any section of a given frequency range (e.g., apply 400 lines from 100 to 200 Hz instead of the 400 lines being dedicated to 0 to 200 Hz). This example would double the frequency resolution.

The first digital FFT was smaller than the time-compression real-time analyzer, but moving all the ancillary equipment around with it was made much easier with a van (Figure 2.29). All it took was a 500-foot reel of coaxial cable!

Figure 2.28   Nicolet 446 FFT in the field.
(Image courtesy of DynaVibes Corporation.)

Figure 2.29   Taking the new FFT to the field.
(Image courtesy of DynaVibes Corporation.)

CASE STUDY

A Canadian power utility was building its own frequency analyzer by writing software to run on a computer. I accompanied one of my firm's salespeople for a demonstration of our new digital FFT, the Nicolet 440. The data were taken from a data recorder at the customer's site and then compared. The customer commented that his plot had many more peaks than the plot from the 440. I thought for a couple of seconds and then asked, "Where are your antialiasing filters?" The customer roared with laughter and replied, "We just ordered them today." See? Basic digital signal-processing training has its usages.

The best way to explain aliased data to laypeople is to remind them what the spoked wheels on a stagecoach look like in a movie. At certain speeds, the wheels appear to be turning backward! This is the result of the movie frames not sampling fast enough to keep up with the forward motion of each individual spoke. The *Nyquist sampling frequency* is the solution. The analyzer must sample the incoming data at least 2.56 times the maximum frequency of interest. In the case of the spokes, instead of sampling that one spoke fast enough as it moves clockwise, the viewer ends up seeing the next spoke as it moves clockwise close to where the spoke of interest was seen. The result is that the viewer's eyes assume that they are seeing the original spoke move slightly counterclockwise instead.

The results of not having antialiasing filters in an FFT analyzer are the folding-in of the higher-frequency peaks into the lower-frequency ranges.

CASE STUDY

We were so proud of the digital FFT! A demonstration at a power utility resulted in data that indicated that the generator was operating at 59 Hz. The lead engineer for the utility scoffed and replied that couldn't be. If that were true, the power grid would kick the plant right off-line. We found out later that analyzer had a bad internal crystal regulator circuit!

The plot in Figure 2.30 was recorded using a handheld FFT analyzer. The result was plotted digitally on an ink jet printer.

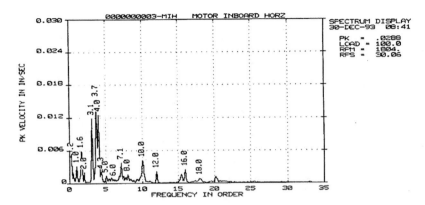

Figure 2.30 Narrow-band FFT of same data as Figure 2.25.

The Entek dataPAC took all the equipment in Figure 2.28 and put it into a handheld box. There was also 10 times the computing capability as well. For example, the FFT could have 12,600 lines of resolution instead of 400. The typical machine requires spectra that are 3,200 lines of resolution to clearly identify faults.

Figure 2.31 Entek dataPAC 1500 portable FFT analyzer.
(Image courtesy of Rockwell Entek.)

The analyzer also contains programs for single- and two-plane balancing and other applications requiring the interpretation of magnitude and phase versus time, such as Bode plots. The dataPAC 1500 also could record data on a preprogrammed schedule. I used this function to find worse-case vibration on a dedicated air compressor at a major helicopter manufacturing facility. These units are still around and are a great value because all the software is included.

State-of-the-art electronics now have taken the size down to what is shown in Figure 2.32. The unit really can't get any smaller or human beings wouldn't be able to view the screen or push the buttons. The photo tachometer is now built into the top of the unit.

Figure 2.32   Enpac 2500.
(Image courtesy of Rockwell Automation.)

Notice the ease of identifying peaks in an FFT. The software, with user input, automatically labels machine speeds and frequencies. The plot in Figure 2.33 identifies fan speed, motor speed, gear pump lobe pass frequency, and other parameters.

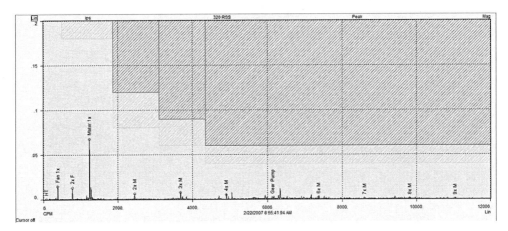

Figure 2.33   Digital plot output from database.

BUILDING A MACHINE FOR PROPER DIAGNOSIS

Building the database to properly diagnose machinery has several key elements:

- List all machines within the complex.
- Separate out the machines that have duplicity.
- Separate out the machines that have a replacement cost of less than a 50-hp motor.

Of the machines that are less than 50 hp, put back into the database the ones that are critical to the operation.

Note: Noncritical motors that are less than 50 hp are deemed replaceable, not repairable. They should be trended for safety issues only.

Now that the technician has a complete list of the machines that require trending, the information concerning the equipment is necessary to build the measurement points and determine what should be measured, including data types and warning and critical alarm levels for each data type. Database programs will also calculate ball bearing fault frequencies from bearing lookup tables included in the software.

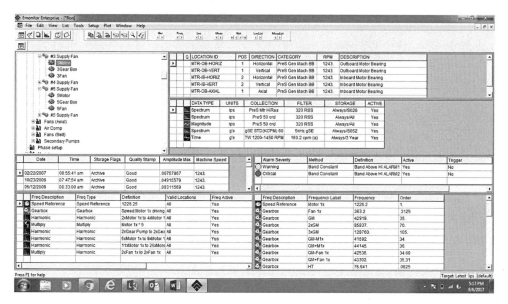

Figure 2.34   Typical vibration-trending database setup for Rockwell Emonitor.
(Image courtesy of Rockwell Automation.)

I want to make a comment concerning the Rockwell Automation Emonitor product. The company has completed the task of automated diagnostics. The mind power of the typical vibration analyst is immediately apparent when one considers that Emonitor requires more than 250 separate pieces of information concerning the makeup of one machine in order to complete the task. This part of the task is equivalent to asking a computerized maintenance management system such as Maximo to do the complete task of scheduling maintenance, ordering spare parts, printing required tool lists and safety instructions, and so on. The software is up to the task, but very few corporations are willing or financially able to put in all the programming time required for the software to do the job.

CASE STUDY

A vibration consulting firm was asked by a manufacturing facility to take over its 10-year-old vibration program. Both firms used the same software package so that the program would not have any discontinuities during the handoff. The database was uploaded to the consulting firm, and the technician started to interrogate the data just to be familiar with what to expect. Every spectrum from every machine had one peak at 60 Hz and nothing else. The initial program setup had an error in the data-collection specifications. There was a small box to check to tell the data-collection box to power the accelerometer. It was not checked! The only data the end user had from the entire 10 years of effort was 60-Hz noise picked up from the cabling.

I would suspect that the program was flawed from the beginning. With proper training and guidance, the technicians would have caught this mistake with their first data-collection route.

The software will automatically add the frequencies in Figure 2.35 to the frequency item list. With frequency labels turned on for a given spectrum, the ball bearing peaks will be identified. Note that the results are order based. To find the frequency, the software multiplies the order times the running speed.

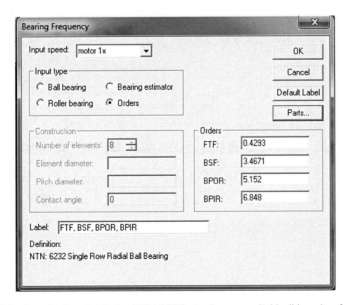

Figure 2.35   Auto lookup table for NTN 6232 single-row radial ball bearing frequencies.

The bearing frequencies are defined as follows:

- FTF: fundamental train frequency—the 1× speed of the cage that supports the balls or rollers.
- BSF: ball spin frequency—the 1× rotational speed of each ball.
- BPFO: ball pass frequency, outer race—generated by each ball passing over a fault in the outer race.
- BPFI: ball pass frequency, inner race—generated by each ball passing over a fault in the inner race.

Do you want to check the math? Add the BPFO and BPFI results, and you should get the number of balls/rollers: 5.152 + 6.848 = 12.

Freq Description	Frequency Label	Frequency	Order	▲
Bearing	FTF	1538.5	.4293	
Bearing	BSF	12425.	3.467	
Bearing	2xBSF	24850.	6.934	
Bearing	BPOR	18463.	5.152	
Bearing	BPIR	24541.	6.848	

Figure 2.36 Example of ball bearing frequencies in orders and frequency based on revolutions per minute.

CALCULATIONS FOR BALL BEARING FREQUENCIES

The formulas used to generate these frequencies are based on the bearing geometries as follows:

Fault	Abbreviation	Equation	Location
Fundamental train frequency	FTF	$F/2 \times [1 - (B/P \times \cos\theta)]$	$<\frac{1}{2}F$
Ball spin frequency	BSF	$P/2B \times F[1 - (B/P \times \cos\theta)^2]$	$5-15F$
Ball pass outer race frequency	BPFO	$N/2 \times F[1 - (B/P \times \cos\theta)]$	$2-15F$
Ball pass inner race frequency	BPFI	$N/2 \times F[1 + (B/P \times \cos\theta)]$	$4-15F$

Where N = number of balls; B = ball diameter; P = pitch diameter; F = shaft speed (use 1 to calculate based on orders of running speed); and θ = ball contact angle.

CALCULATING AND IDENTIFYING BELT FREQUENCIES

Technicians will encounter many machines that are belt driven. The formula for calculating the frequency at which the belt(s) travel over each sheave plus the distance between them is calculated as follows:

Belt frequency = π × sheave rpm × sheave pitch diameter/belt length

So, for a sheave that is 10 inches in diameter, turning at 1,800 rpm, and driving a 70-inch belt,

3.14159 × 10 × 1,800/70 = 808 rpm

Of course, one would have to know these dimensions ahead of time. This is never the case. Belts are also notorious for generating harmonics. This makes them very easy to ferret out in the spectrum, unless the motor is operating on a *variable-frequency drive* (VFD). ***The analyst is warned ahead of time: get the drive frequency entered manually into the database* before *you leave the machine.*** It is next to impossible to figure out the frequencies after the fact.

Create a separate data-collection point in the measurement for *VFD fundamental frequency*. The technician can enter this one time before taking the data, and all readings can then be referenced to this value, automatically lining up data, frequency labels, and so on.

RECOMMENDED MEASUREMENTS FOR ANALYSIS

The field technician usually has two separate tasks when taking vibration on a machine in the field. Regularly scheduled routes mean taking hundreds of pieces of information in as short a time span as possible. The other task involves diagnosing a problem. Each task has unique measurement sets. The trend capture is meant to find the machines that are no longer operating within satisfactory limits.

TYPICAL MEASUREMENT SETUP FOR ONE-TIME DIAGNOSTICS

I have used the settings in Figure 2.37 successfully.

	DATA TYPE	JNITS	COLLECTION	FILTER	STORAGE	ACTIVE
▶ ∿	Magnitude	ips	PreS Mtr HiRes	320 RSS	Always/S026	Yes
꜒꜒	Spectrum	ips	PreS Mtr HiRes	320 RSS	Always/All	Yes
∿	Magnitude	ips	PreS 50 ord	320 RSS	Always/S026	Yes
꜒꜒	Spectrum	ips	PreS 50 ord	320 RSS	Always/All	Yes
∿	Magnitude	ips	PreS Mtr HiFreq	320 RSS	Always/All	Yes
꜒꜒	Spectrum	ips	PreS Mtr HiFreq	320 RSS	Always/All	Yes
∿	Magnitude	g's	PSgSE(KCPM)60	5kHz gSE	Always/All	Yes
꜒꜒	Spectrum	g's	PSgSE(KCPM)60	5kHz gSE	Always/All	Yes
∿	Time	g's	TW 0920-1313 RPM	162 RSS	Always/All	Yes

Figure 2.37   Measurements for diagnostics.

Note: Most data collectors take time to change filter settings for each reading. Readers can see that the "FILTER" settings have been grouped to take all similar readings together to speed up data acquisition.

All software will have the ability to select the same settings as in Figure 2.37. They include magnitudes, which are beneficial for trends but are also valuable for multiple readings during a troubleshooting session.

Spectrum Settings

Frequency range:	0–12,000 cpm, 3,200 lines of resolution (PreS Mtr HiRes)
	0–50 orders of running speed, 3,200 lines of resolution (PreS 50 ord)
	0–300,000 cpm, 3,200 lines of resolution (PreS Mtr HiFreq)
Demodulated spectrum:	0–60,000 cpm filtered out, 300,000 maximum frequency (PSgSE)
Time waveform:	Calculated time base to include *at least* six full revolutions of the fundamental.

There are so few of these collections required that the technician can build them ahead of time.

Take the most common speeds encountered in the field, say a two-pole motor speed of 3,600 rpm. The time it takes for one revolution of a two-pole motor is 1/60 Hz = 0.01666 s. Time for six revolutions is 0.1 s.

The time collection in a data collector is a binary number. For example, 2,048 time points become an 800-line-resolution FFT because of the Nyquist frequency: 2,048/2.56 = 800 lines.

Therefore, we can select a 7,000-Hz F_{max} with 800 lines of resolution. This will yield a time waveform with 2,048 time points for a total time collection of 0.11423 s.

Now the technician can create numerous collections at ease:

- A four-pole motor requires an Fmax of 3,500 Hz.
- A six-pole motor requires an Fmax of 1,750 Hz.

The belt fundamental frequency will always be lower than both the driver and driven fundamental frequencies. The spectrum in Figure 2.38 has two peaks to the left of the fan speed. Selecting the first peak with the harmonic cursors on identifies three other peaks that are harmonics of the fundamental.

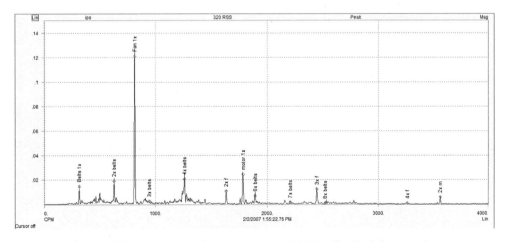

Figure 2.38 Example of four-pole motor with identified belt frequencies.

A predominant peak much lower in frequency than either fan or motor speed is identified. The cursor is placed on that peak, and then the harmonic markers are turned on. Boom! Multiple peaks are identified as harmonics of the belts (black dots in Figure 2.39). Now we just have to identify motor 1× and fan 1×. Again, unfortunately for today's analyst, safety restrictions have all but made the sheaves and belts invisible, completely covered. Comparing the sizes of the two sheaves, the technician can make certain assumptions. The shaft with the larger of the two sheaves will have a lower 1× running speed. Also, viewing the vibration readings from the motor and fan bearings should make this easy to ascertain because the fan vibration should be more prevalent on the bearing furthest from the sheave. The bearing closest to the sheave sometimes can exhibit a high motor level from transmission through the belts.

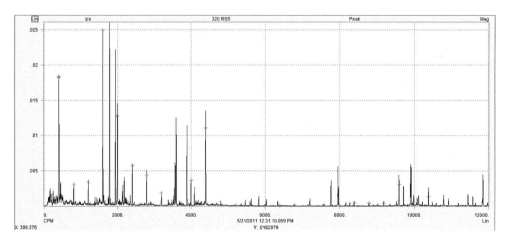

Figure 2.39   Example of finding belt harmonics.

Once the belt frequencies are identified, there are two major peaks near what would have to be the fan and motor fundamental frequencies. One of the peaks was higher than 1,785 rpm: the base speed of the motor. Using variable-speed drives to overspeed motors is not common, so it is assumed this was the fan speed. The other peak is labeled as "motor 1×." This particular machine does have the larger sheave on the motor, giving the fan a higher speed. Once the database has been set up with the criteria in Figure 2.40, automated reports can be generated, saving time.

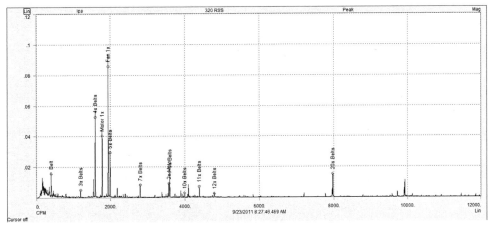

Figure 2.40   All frequencies identified.

C A S T   S T U D Y

A customer had built an extensive database to analyze a large paper machine. This machine consisted of hundreds of rolls and other geared items. This was one of those machines that seldom run at the same speed, so how does the database handle this? The customer normally referred to the machine speed by the paper speed in feet per minute. This was input by the vibration technician at the start of their route. This input also can be fed into the handheld data collector from a speed reference sensor if speeds change between data-acquisition points. The data are all recorded and dumped back into the database. The database then can take that speed reference and calculate the speed of every other point tested by knowing the speed ratios. The data are all automatically correlated for speed.

Note: A hardwired speed reference input to the data collector is recommended in case there are any changes in speed between data-collection points.

Completeness of all the above-mentioned information will result in faster response times for the customer concerning the present condition of the machinery.

Notice in the plot in Figure 2.41 that the motor vibration frequencies are clearly labeled, as are the fan frequencies and the oil gear lube pump on the gearbox. The warning and critical alarm bands are also illustrated, which confirm that there are no issues on this piece of equipment at the present time. Most vibration database programs can automate a report based on any peak crossing the alarm limit. They also can sum the energy in any given band. This is important because a large amount of vibration energy could be damaging, even if there is no single peak that triggers the alarm limit. These alarms are used in the higher-frequency bands—for example, where gear mesh modulation and antifriction bearing energies reside.

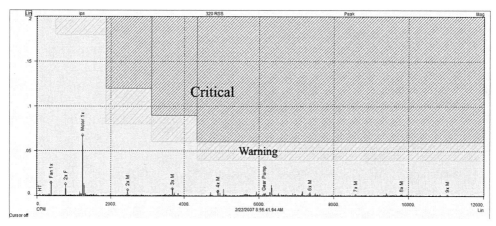

Figure 2.41   Velocity spectrum with key machinery components labeled.

Every frequency is calculated from the "speed reference." Therefore, variable-speed machinery should have the speed reference captured by a hardware tachometer.

ALARM BANDS FOR VIBRATION-TRENDED DATA

Customers who have multiple readings archived require additional tools to interrogate data. Multiple readings are saved, including magnitude, spectra, time waveforms, and so on. Alarms are not only an immediate visual assistant, but they also can be part of an automated report (Figure 2.47). The report can reduce the amount of data the technician needs to view by as much as 90%.

The spectrum would have alerted the technician to a critical band alarm excursion at not only 1× but also 2× line frequency (Figure 2.43). An automated report also should be created that will include the plot, along with a complete description of the machinery point.

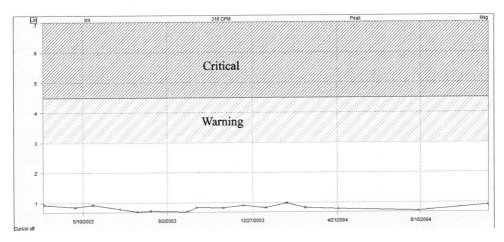

Figure 2.42 Overall velocity magnitude trend with warning and critical magnitude alarms.

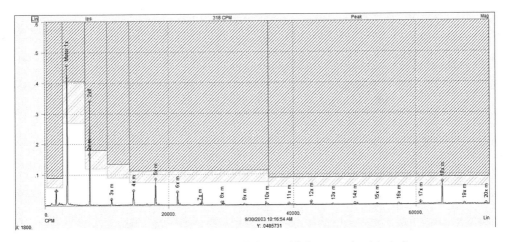

Figure 2.43 Spectral band alarm with frequencies labeled.

The spectral alarms are beneficial when multiple readings are available, either from one machine or from a family of similar machines. The user can select the data and build the alarm. Then the selected reading on a route can be compared against a statistical average at all frequencies for all machines (Figure 2.44). Notice that the 1× and 2× running speed vibrations for this particular pump at this particular time are almost 50% lower than average.

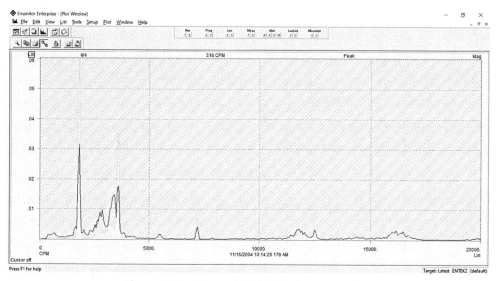

Figure 2.44   Spectral statistical alarm—warning.

Alarm bands are flexible and can include expected higher harmonic peaks, such as vane pass frequency generated from an impeller with seven vanes (Figure 2.45).

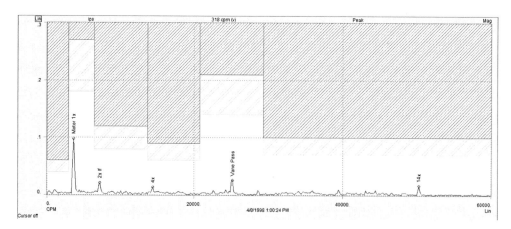

Figure 2.45   Spectral alarm band with vane pass frequency included.

Portable FFT analyzers can take vibration data, temperature data, and so on and convert the incoming accelerometer data into various formats. All these data should be viewed to make the best assessment of the machinery condition (Figure 2.46).

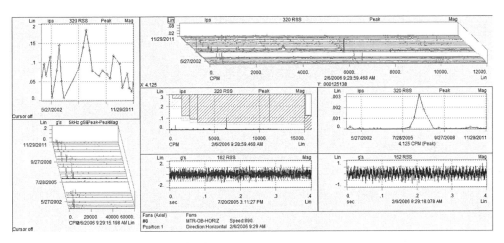

Figure 2.46 Viewing all recorded data.

Some database products—Emonitor, for example—have their data displays hyperlinked. This allows the user to move the cursor on any one of the data windows and all the other windows will update to that date (in the waterfall plot) or that frequency. The data window also can include a thermographic image, again from the hyperlinking ability.

CASE STUDY

A large manufacturing firm had more than 1,500 points on a permanent monitoring system. The computers would sample the data and store them once every shift. The technicians would arrive for the day shift and start to go through the data, plot by plot. They could analyze approximately 50% of the previously stored data, when the latest shift data would arrive in the database! Obviously, this is not ideal. Corporate leadership was prepared to abandon the project as worthless. An inspection of the database software uncovered the fact that the original system installer had not set up any alarm bands or reports. The customer wasn't utilizing the total power of the database software.

The historical data were important to the customer. As a result, it took two technicians 12–18 days, over a three-month period, to add the alarms and reports, being careful not to disturb or overwrite any historical data. The final product was three reports that were ready before the day shift started. One report tagged any transducers that were faulty, the second report identified any point that alarmed on spectrum velocity alarm bands, and the third report notified the techs of any point where the acceleration-time waveform was in alarm. The total printout of all three reports was approximately 5–10 pages (Figure 2.47).

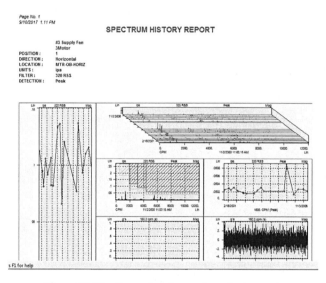

Figure 2.47 Example of an automated report.

The technicians also were instructed to generate a report that totaled saved downtime as a result of their work. The downtime costs were calculated by the accounting department. This is important! The costs must come from the people responsible for determining profitability of the facility. Month after month, savings were in excess of $50,000 to $80,000. These costs were on the conservative side, not even considering replacement costs for machinery damaged as a result of running at excessive levels. Even worse, some parts might not be readily accessible with long lead times. This was all accomplished while reducing the technician's analysis time by 75%. They were now free to perform proactive maintenance items, balancing, alignment, and so on.

The vibration severity chart in Figure 2.48 was created by PMC/Beta Engineering. The chart uses velocity readings to characterize machinery overall health. Notice that there is a roll-off on the low-frequency end, below 1,000 rpm. Also, there are multipliers for various types of machinery, which by their operation nature may have higher normal vibration levels.

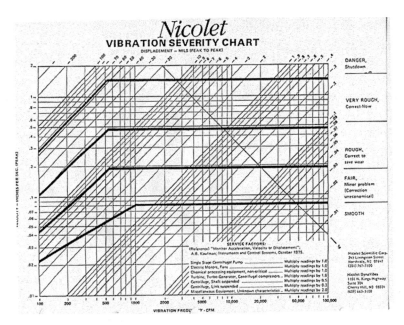

Figure 2.48   Realistic vibration severity chart.
(Image courtesy of PMC/Beta.)

When putting together a large machinery database, take into consideration how a family of machines is behaving. If 10 similar motor/pump applications are all within a "fair" vibration range but two others are in the "rough" category, it's worth the time to diagnose what is causing the elevated levels to make certain that the machines are electrically and mechanically without faults. Also check process variables against designed criteria.

REVIEW QUESTIONS

1. A pure sine wave has a maximum peak value of 1.2 V. What is the area under the curve?

2. Displacement readings are always captured as peak-to-peak values. What would be the displacement value in mils peak to peak of the previous reading with a 0.200 volt/mil sensor?

3. True or False: The RMS can be calculated for a nonperiodic time waveform.

4. True or false: RMS stands for *relative mean squared*.

5. True or false: Zero to peak and peak readings are the same thing.

6. True or false: Peak-to-peak readings are used for displacement only.

7. True or false: The relationship between displacement, velocity, and acceleration is linear.

8. Fill in the missing parameter: Velocity $= ? \times f \times D$ (where f = frequency in Hz, D = displacement in inches peak to peak).

9. What is the value of g in English units?

10. What do digital spectrum analyzers use to prevent the low-frequency ski slope?

11. What is a beat frequency?

12. Does a beat frequency increase or decrease as the frequencies get closer to each other?

13. What is the beat frequency of two motors, side by side, at 3,585 and 3,580 rpm, respectively?

14. How many times a minute will a technician "feel" this beat rise and subside?

15. Power plants need extremely accurate speeds on their turbine generators. One customer had two generators running side by side, and the time between maximum beats was 2 hours. Calculate the beat frequency for these two units.

16. Give a list of descriptive labels for the bearings of a motor-driven fan.

17. Because both radial readings on the top of a vertical motor are horizontal, how can you differentiate them?

18. Why is there only one axial reading on a given machine?

19. Two sine waves that are out of phase could still have the same two parameters. What are they?

20. What is the major drawback of analog-frequency analyzers?

21. What is the minimum frequency analysis rate for a given maximum frequency range of interest?

22. What is the minimum horsepower for trending to extend the life of the machine if it's noncritical?

23. What is the answer if you add the ball pass frequencies of the inner and outer races?

24. What is the most important vibration parameter for monitoring machinery overall health?

25. What key parameter in the data collector and/or route setup must be checked to ensure proper collection of data with an accelerometer?

26. Motor speed and fan speed are both higher in frequency than what on a belt-driven machine? What is the formula for calculating belt frequency?

27. Time waveform collections should have settings that ensure how many full cycles of 1× running speed?

28. What is the best analysis path for breaking down a spectrum with a large number of peaks?

29. What critical piece of information is required to post analyze variable-speed machinery?

30. Report writing is very time-consuming. What should be added to a trending database to quickly interrogate vibration data yet yield a quality report?

31. What alarm type can assist in differentiating between a "good" and a "bad" machine where there are multiple machines of similar design and operation (for example, six pumps of similar horsepower and speed)?

32. What is the most valuable commodity from a trending program that must be communicated to management of a given facility?

CHAPTER 3

Trend Programs

Trending programs are very beneficial for reducing downtime, thereby saving companies money. Most of the savings are captured through the reduction of downtime. Additional benefits shouldn't be minimized, such as preventing accidents that could cause personal injury. The best practices can save a machine from a catastrophic failure of a gearbox that may have a replacement lead time of 6 months or a redesign because of obsolete parts. Vibration trending programs are meant to isolate any machine not operating at satisfactory limits. For that reason, the data collections can be more limited in scope.

Now let's consider a technician who has been asked to look at 250 machines over the time span on one day. Many customers ask for trending analysis quotes based on a price per point. I have found this practice to be meaningless because it isn't fair to the customer or the technician. A fair cost is based on the amount of time it takes the technician to collect the data, analyze it, *and* report the results. There also should be caveats concerning the availability of the equipment. A technician can't be expected to wait a long period of time for a machine to go online.

The following is an example quote for vibration analysis. The customer was 90 miles from the shop. All data collection had to be done on the shift.

Proposal for Vibration Analysis on the Following Equipment

30 Rolls
4 Gearboxes
4 Motors
8 Motor-driven fans
6 Motor-driven pumps

Total price for the initial vibration analysis is $2,500. This initial visit also includes the database build. The database will be completed with accurate measurement locations and component information based on input from the customer.

Bimonthly vibration trending: $1,250. Price includes all travel, mileage, and tolls; the vibration analysis; and a full report.

Scheduling will be coordinated with the customer to ensure that equipment is in operation. Any equipment that is not operating will be examined on the next scheduled visit or the data can be collected on an additional day at our regularly scheduled field-service rate.

The final report will include any discrepancies found and recommendations for fixing the problem in a cost-effective manner.

The following is an example that required some thinking outside the box. The customer really didn't know the full extent of the route on the initial visit. There were also concerns about critical frequencies existing at certain speeds on the pumps and cooling towers once the variable-frequency drives (VFDs) were installed and running.

Proposal for Vibration Analysis on the Following Equipment

6 Supply fans
2 High-velocity fans
2 Return fans
11 Pumps
2 Cooling tower motors

If there are any discrepancies in the total number of machines or descriptions, please advise. Any additional equipment added to this list that is eight pieces or less will not change the scope or price of this proposal.

The total price for the initial vibration analysis is $2,500.

Follow-up for adjustable-speed resonance check: $1,250.

The price includes all travel, mileage, and tolls; the vibration analysis; and a full report.

A critical speed analysis can be performed on the chill water pumps and the cooling tower motors after the drives are installed and the speeds can be varied. The initial analysis will identify any speeds that should be avoided because of excitation of natural frequencies that could cause resonance. The response will be of a higher magnitude once the machine can stay at one speed in the operating range. For this reason, we recommend a follow-up visit once the drives are installed.

An example of a report page is included in this proposal. The final report will include any discrepancies found and recommendations for fixing the problem in a cost-effective manner.

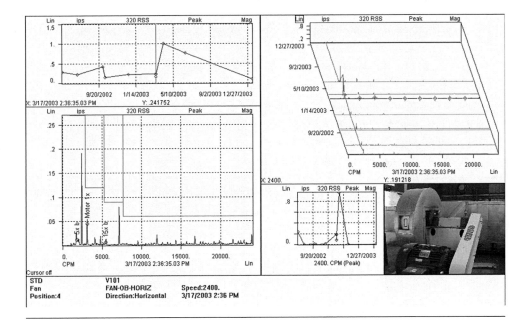

The price for the critical speed analysis is low because of the amount of work required. A peak hold spectrum is captured while the machine is taken to full speed and then put in a coast-down. There might not be any resonance points or maybe one frequency range issue. Note the frequency on the plot, block out a 15% band around that speed, and note it on a report to the customer.

CASE STUDY

The fan in Figure 3.1 had been installed with four-pole motors and gear reducers to run at a fixed speed of 450 rpm. The customer installed VFDs on the motors to better regulate air requirements and save energy dollars. The outboard fan-bearing pedestals were 10 ft high, manufactured from structural steel anchored into the concrete floors on the 19th floor of the building. Building personnel witnessed high vibration on the outboard pedestal with the fan operating at certain speeds. A local firm called for dynamic balancing of all five fans. The customer asked for a second opinion.

The pedestal vibration on the center hung fan was predominantly in the axial direction of the outboard fan pedestal. It was not a balance problem at all. The plot clearly indicates that vibration at full speed was very low (right side of chart). Vibration due to imbalance will always increase with speed, increasing at a rate of a square of the speed change. Double the speed, and the vibration will increase by a factor of 4. Again, this was *not* an imbalance. The critical frequency was identified at 338 rpm. Notice that all other measured points did not reflect the fan pedestal axial resonance. The recommendation to the customer was a dead-band span of 15% centered on this frequency. The rule of thumb for avoiding natural frequencies is a separation of *at least* 5%. This dead band guarantees a separation of 7.5% above and below the resonance point.

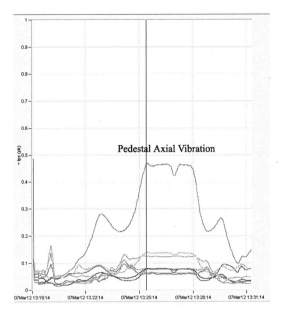

Figure 3.1 Vibration analysis and report.

MEASUREMENTS FOR TRENDING PROGRAMS

A vibration trending program has a specific task: find the machines with vibration levels that are above satisfactory or increasing with time. For this reason, the measurements can be abbreviated from the set used for diagnostic purposes. I created two templates: fast fan and fast pump. These cover 90% of the machinery trended for customers. The addition of a gearbox or something else will call for additional readings: for example, a velocity spectrum with 200 orders and 800 lines.

The measurements were as follows:

Motor OB Horizontal
50-order velocity spectrum, 800 lines of resolution
5,000-Hz velocity spectrum, 800 lines of resolution
200-Hz velocity spectrum, 3,200 lines of resolution
6 revolution time waveform, 800 time points
Demodulated spectrum, 800 lines of resolution

Note: The user is encouraged to use overlap processing to increase data-collection speed. The analyzer will collect 1,600 time points for an 800-line spectrum. Fifty percent overlap processing will keep 800 of those time points and collect only an additional 800 time points for the next average. This is a great time savings. Averaging will smooth out vibration amplitudes over time, but most of the vibration encountered in pumps and fans will be periodic and stationary. The reader is encouraged to run an experiment. Make two copies of this template. Change the averages from two to eight on the copy. Also make the overlap processing 50% on the two-averages template and 0% on the eight-averages template. The differences in the data should be negligible.

There are two measurements necessary for electric motors, the 5,000-Hz, 800-line spectrum and the 200-Hz, 3,200-line spectrum. One each should be taken on the motor horizontal locations.

The 5,000-Hz velocity spectrum reading is for identifying any modulation of the rotor in the field. This will cause rotor bar pass frequency with multiple ±2× line frequency sidebands. This is not indicative of broken rotor bars.

The 200-Hz velocity spectrum has sufficient resolution to capture pole pass sidebands surrounding 1×, 2×, and 3× running speed and 2× line frequency. Pole pass frequencies are calculated as the difference between the field speed and the rotor speed, or slip frequency, times the number of poles. For example, a two-pole motor has a field speed of 3,600 cycles per minute and a rotor speed of 3,585 revolutions per minute:

3,600 – 3,585 = 15 cpm × 2 poles = 2 × 15 = 30 cpm

A rotor with broken bars or high porosity in the end rings will generate ± 30-cpm sidebands. Without 3,200 lines of resolution, this will not be distinguishable (see Figures 3.2 through 3.4).

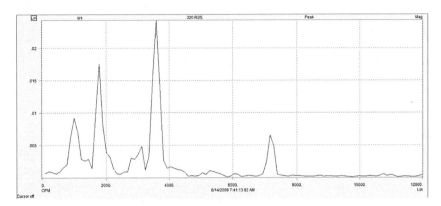

Figure 3.2   A 50-Order, 800-line spectrum expanded to look for rotor condition.
The 800-line spectrum does not have enough resolution to assess the condition of the rotor.

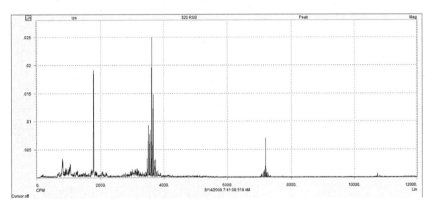

Figure 3.3   A 12,000 CPM, 3,200-line spectrum of a two-pole chiller motor with a bad rotor.

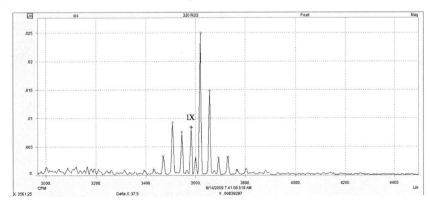

Figure 3.4   Running speed with 2× slip sidebands.

This is indicative of a bad induction motor rotor. The "1X" text is positioned at running speed, 3,581.25 rpm. The sideband cursors are reading ±37.5 cpm. Notice that this is twice the difference between 3,600 cpm and 3,581.25 rpm.

The horizontal readings on the motors and fans contain most of the data. The vertical and axial readings are as follows:

MOV and MOA
Velocity magnitude trend
50-Order velocity spectrum, 800 lines of resolution
Demodulated spectrum, 800 lines of resolution

The measurements are the same at the MIV and MIA locations, not needing the high-frequency and high-resolution readings. The only difference between the fan and pump templates is the alarm band for vane pass frequency, which is more prevalent in pumps.

CASE STUDY

Consider this example of a piece of equipment that was part of a trending program, as shown in Figure 3.5. This machine has a vertically mounted motor with a close-coupled fit to a multistage gearbox. The motor has a speed of 1,800 rpm. The impeller speed is approximately 60 rpm. This means a gearbox with a lot of gears (see Figure 3.6)!

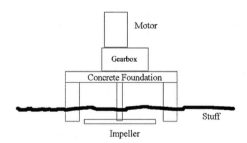

Figure 3.5 Simplified drawing of motor and gearbox agitator.

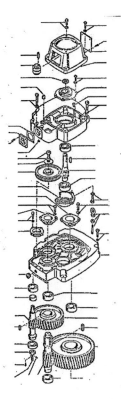

Figure 3.6 Exploded view of gearbox.

The complexity of this gearbox is evident from first glance. The need to categorize all the gear teeth, bearings, and so on would take a long time. The machine was still successfully trended for years.

The baseline signature for this gearbox was unique and not like any of the other seven gearboxes (Figure 3.7). The vibration, centered at 80,000 cpm, was of the most concern because it was probably gear noise.

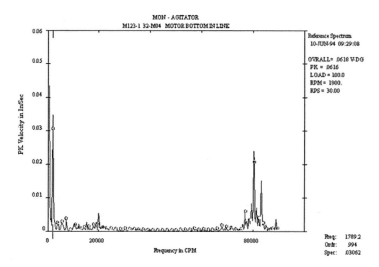

Figure 3.7   Baseline signature on gearbox.

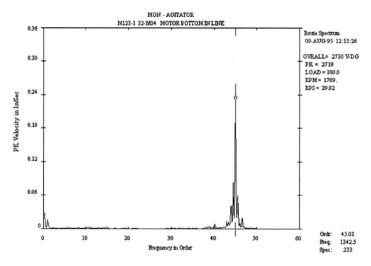

Figure 3.8   Velocity spectrum at the highest point in the trend.

The time waveform modulating is displayed in Figure 3.9. This is the result of one higher frequency riding on a lower frequency. The higher frequency is gear mesh, or number of teeth times running speed of the gear. The lower frequency in the time waveform is 1,800 rpm. This indicates that the gear having issues is the first gear in, or the gear attached to the motor input shaft.

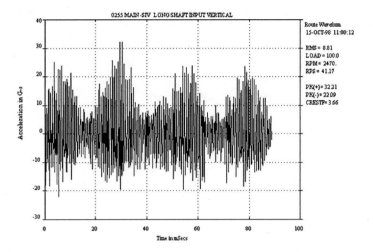

Figure 3.9   Modulation of the first gear set.

This trend identified an increase in overall vibration on the motor–gearbox connection point. The drop-off in the vibration marks the date when the problem was identified and then resolved (Figure 3.10).

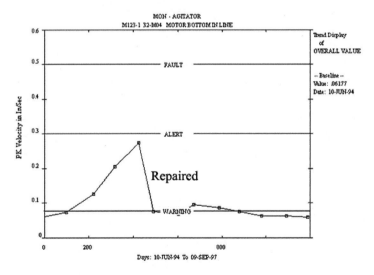

Figure 3.10   Overall velocity trend on lower motor bearing.

The technician can use alarm bands to further quantify the gathered vibration data for diagnosing a fault. Alarm bands can be categorized as peak or sum. A *peak* band will alarm if any one frequency bin crosses the threshold of the alarm. The *sum* bands are best for that area of the spectrum where the user can expect broadband energy from bearings and/or gear noise. This example used a sum band for the gear application, as shown in Figure 3.11.

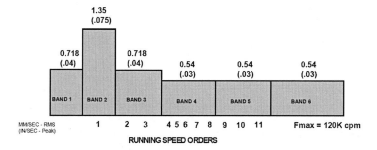

GEAR BOX
LINE AMPLITUDE ACCEPTANCE LIMITS

Figure 3.11   Full frequency range divided into bands for gear analysis and trending. (Courtesy of General Motors Vibration Standards.)

The rapid increase in vibration within this band is evident. There was something wrong with the gearing. The beginning of this trend occurred just after the maintenance people had replaced the motor (Figures 3.12 and 3.13). The job was performed during a blizzard on the shore of a major river in the Northeast in the United States. The company had pulled the motor down into a close-coupled fit of the upper gear–bearing assembly with the key cocked. The result was that the top bearing set had spun in its housings after a short period of operation. The loose bearing caused a modulating vibration in the top gear set.

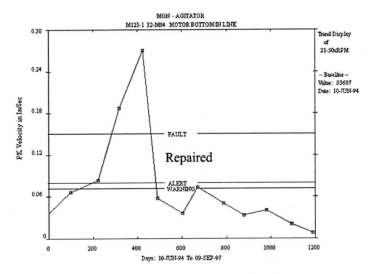

Figure 3.12   Band 6 for gear applications.

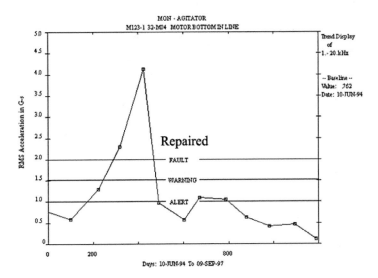

Figure 3.13   Acceleration trend.

Based on the vibration technician's recommendations, the company pulled the motor and disassembled just the top bearing–gear set and brackets and sent them to the motor shop. The bearing brackets were bored, sleeved, and brought back to specifications. New bearings were installed on the gear shaft. The keyway on the motor shaft was cleaned up, and the machine was reassembled. Because of complete information, the very cost-effective repair was successful, saving maintenance dollars.

Note: All data have been included on the following case history. Put on your thinking cap.

A customer with an existing vibration trending program asked for an evaluation of its data after missing the catastrophic failure of a 600-hp motor-driven gasoline pump. The customer was using the same software as the motor repair shop, so it sent a template containing all the data from that machine. The trend was almost 2 years in length.

The very first plot viewed by the shop technician identified a severe problem. Why wasn't this problem caught during data analysis?

Earlier in this chapter it was noted that technicians should look at *all* the data. In this case, the technician had not recorded time data (lower center and right in Figure 3.14). The data that were recorded identified a serious problem that had been months in the making. The overall spectral map and the 1× slice up through that map identified the 1× vibration on a continuous and steady increase. The overall acceleration trend in the upper-left plot indicated impacting that had grown rapidly over the last three data collections. But most important were data captured in the acceleration spectral map (lower-left plot)! The pump bearings had been failing for months (Figure 3.15).

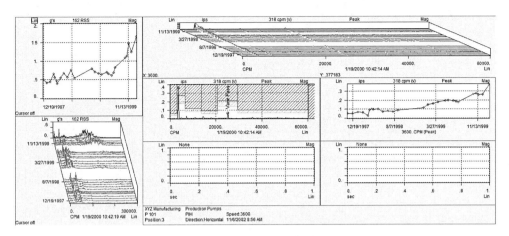

Figure 3.14   Two-year vibration trend on motor–pump assembly.

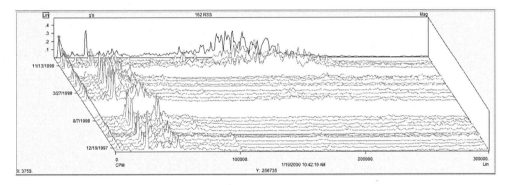

Figure 3.15   Antifriction bearing failure.

Why had the bearing failed? Was the answer in the database?

The steady increase in 1× motor vibration (MOH), the increase in *g* force in the bearing, and a failed motor bearing were indicated in the last data collection (see Figure 3.16). Acceleration spectra had been set up for quarterly readings only. Flexibility in trending software allows for the user to set the time interval for collecting various parameters. This can save time on the data-collection route.

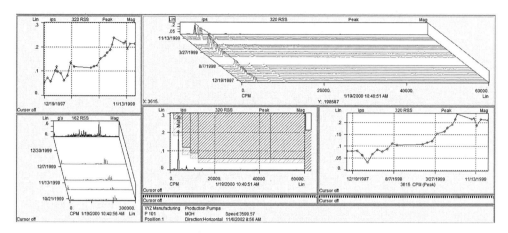

Figure 3.16   Motor outboard horizontal.

Most trending programs don't capture the same amount of data in all directions, so this was typical. However, the data did indicate something important (Figure 3.17). The overall velocity plot (upper left) in Figure 3.17 indicates an increase over time to a final reading over 0.15 inch/s. The spectrum waterfall plot (upper right) and the 1× trend under it don't show this increase. There was something higher in frequency that wasn't captured. The maximum frequency range for the spectrum was too low.

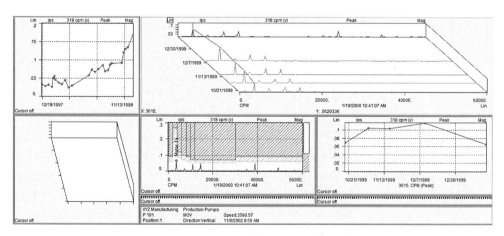

Figure 3.17   Motor outboard vertical.

The axial vibration on the motor holds a key piece of information (Figure 3.18). Four months before the failure, the axial 1× vibration started a downward trend! This trend is very similar to the motor outboard horizontal readings (Figure 3.16).

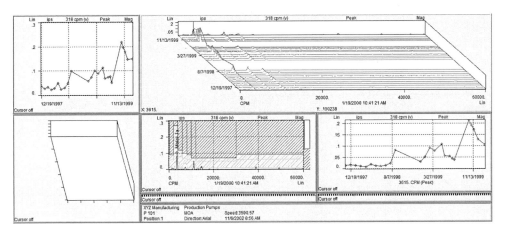

Figure 3.18   Motor axial.

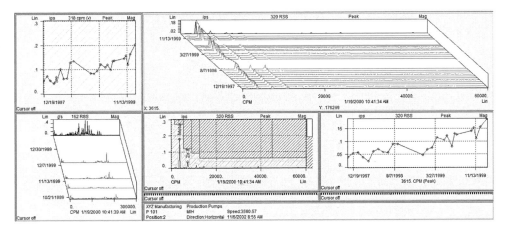

Figure 3.19   Motor inboard horizontal.

Again, there was a sharp increase in the velocity, but an event 4 months before the failure changed the trend to a downward direction (Figure 3.20)! The axial vibration data do not indicate an identifiable fault (Figure 3.21).

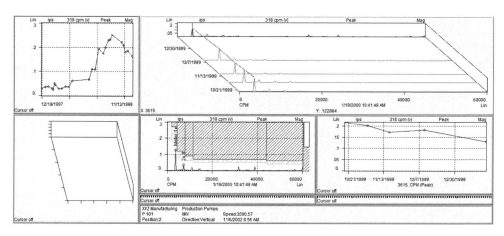

Figure 3.20   Motor inboard vertical.

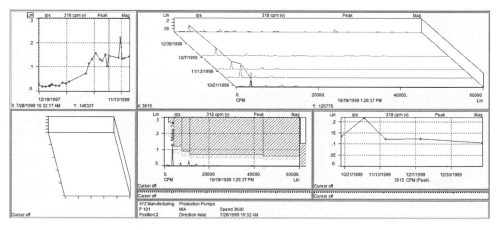

Figure 3.21   Motor inboard axial.

The pump inboard bearing radial readings both show the same continuous trend up to failure (Figure 3.22).

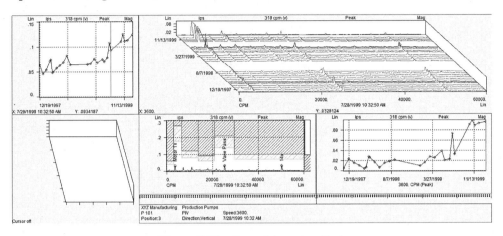

Figure 3.22 Pump inboard vertical.

The pump axial reading identified the same event as the motor axial reading. The spectral map and the 1× speed trend both indicate the problem was something that had a large effect on the 1× operating speed in the axial direction (Figure 3.23). It started to drop!

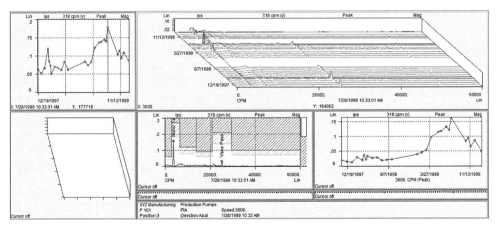

Figure 3.23 Pump inboard axial.

The pump bearing failure is evident in Figure 3.24. This also shows the highest vibration amplitudes encountered in excess of 0.4 inch/s.

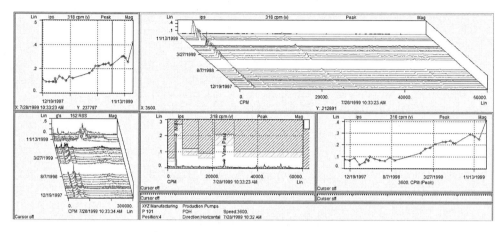

Figure 3.24 Pump outboard horizontal.

These were the same indications as the other pump radial readings (Figures 3.24 and 3.25). The same event that happened approximately four months before the failure is evident in the pump axial 1× running-speed data (Figure 3.26).

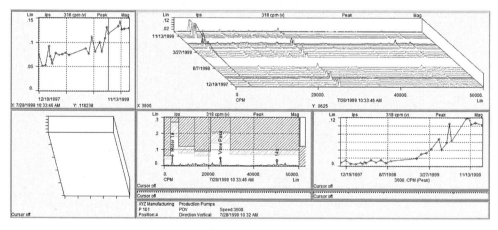

Figure 3.25 Pump outboard vertical.

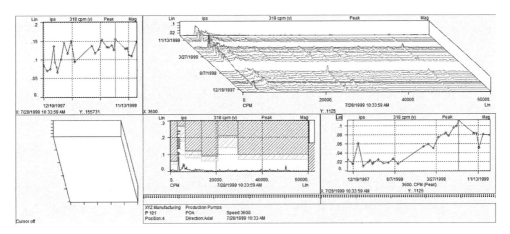

Figure 3.26   Pump outboard axial.

Coming up with a root cause for the failure for this machine was extremely difficult. The motor arrived at the shop in eight separate crates. The pump and piping were a mangled mess.

Based on all the vibration data available and knowing the configuration of this motor–pump assembly, the most probable cause was a failure of the pump piping supports. There is a period of missing data near the event 4 months earlier. When the data were again recorded, the pump bearings were already failing. The piping strain on the pump caused a misalignment of the pump with the motor. The motor had sleeve bearings, and the pump had ball bearings. The excessive misalignment caused the pump bearings to fail prematurely, and that failure and the misalignment caused the motor bearings to rub and fail.

What was the reason this failure was missed? The software was paid for and installed, but the reliability engineers weren't given the time or didn't take the time to install alarms and automated reports (the alarms on the plots were added by the author after the failure). As a result, the customer's vibration technicians had the time to take the data but no time to interrogate the data.

I was part of one of the first predictive routes at a major oil refinery. The maintenance manager asked for a meeting where he announced that the program was being canceled. The participants in the meeting were dumbfounded. We were very proud of having addressed multiple issues each month, and our reporting had concisely listed them, including recommended actions to remedy the situations. The maintenance manager smiled and commended us on our work: "You don't understand. This program isn't being canceled because of anything you are doing wrong. What good is finding all these problems if my people don't act on them?" It is imperative that factual downtime savings are attached to a predictive maintenance program.

This instance involves the same refinery. A pump was analyzed by my technician on the day of installation. The next morning, the refinery's reliability engineer called me to complain that the pump had failed, and he demanded to know why my team had missed the call. I frantically culled the data for a problem, but the data looked fine. I asked the engineer what was found when the pump was brought into the repair shop at the refinery. He hesitated and then replied the failed pump couldn't be found! It had disappeared. He hesitated again and then hung up, saying, "Sorry, never mind."

CASE STUDY

Here's another instance at the same refinery. High vibration was observed on a 100-hp, 3,600 rpm pump. The vibration readings indicated bad misalignment. The reliability engineer called the conclusion "BS." He continued, "That motor was just replaced and aligned last night." Then he headed out to the pump with his alignment tool set, dial indicators, and brackets. I was present when the coupling guard was removed. The first question we all asked ourselves was, "How the heck did the technicians manage to bolt up that coupling?" Just for grins, the brackets and dials were mounted, and readings were taken. The misalignment, a record as far as I am concerned, was 0.750-inch vertical misalignment. My team had to keep record of the total revolutions of the dial because it was calibrated for 0–0.100 inch. Yup. Seven and a half revolutions. The replacement motor wasn't the correct frame size for the pump, so the motor was ¾ inch high. All I could think about was that the reliability engineer's face was within inches of that machine while he took the readings earlier.

Again, another problem at the same refinery. The maintenance manager asked for vibration readings on a fan that was supported by metal framework some three stories high. The vibration readings were again a record, 230 mils! The process had overheated the fan shaft and bowed it. My company presented the maintenance manager with a plaque for the highest vibration ever recorded. He kept it on his wall until the day he retired.

CASE STUDY

The same refinery had another problem. I was present to balance a centrifuge at 3 a.m. The balancing procedure was completed, and vibration readings were taken periodically. Suddenly, a large bang was heard from up in the cracking tower. It was repeated by three larger bangs, and then it stopped. I ran into the control room to alert the technicians of the problem. They came outside, heard nothing, and went back in. This banging was repeated several times every hour. Then the sun started to rise, and the problem was visible. There was a 24-inch water supply line up in the tower that was moving a foot with each bang! Still, there was no action from maintenance. Then, at about 7 a.m., the line ruptured. The resulting waterfall was picturesque.

Many of you must be asking yourselves about now, "Why the heck did you go to that refinery?" My company was young and hungry. You go where the money is. There has been many a job in my career where things learned with a quick analysis on that job site paid off.

CASE STUDY

Another customer had a third party documenting the vibration routes and writing the reports. The customer insisted on a printout of *all* the data. While I was there to address a separate issue, the customer wanted to know if my firm would want to take over the vibration route. He was concerned that his third-party regular was missing problems. For example, the company had recently lost a pump without warning. I grabbed the latest huge binder and looked up the pump on the first-page summary. There, under the pump listing, was one sentence, "Change pump bearings ASAP." *Never* give a customer all the data if you can avoid it. The report should consist of a short list of only the equipment with problems. Then it needs to be read and acted on.

I had only one customer that demanded all the data. After 1 year on the program, the binders had filled one whole shelf in his bookcase, and he pleaded with me to stop sending all the data! Each report contained one or two pages in the front that listed the machines in need of servicing. After I downsized the data for this customer, the entire year fit into a 1-inch binder. This binder included all vibration plots required to justify the call for action in the recommendations.

Figure 3.27 illustrates the benefits of using a vibration trending software with hyperlinked programming. The image on the lower right is the gearbox of a large cooling tower. This image was taken right through the protective screen with the fan in operation. Any plot from any software package can be included in the page—for example, a photograph of the machine, thermographic image, and so on.

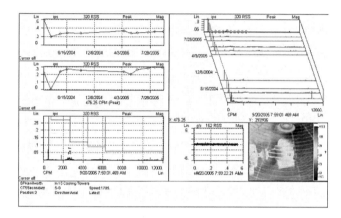

Figure 3.27   Combined vibration and thermography report plot.

THERMOGRAPHY

Thermography is an important tool for proactive maintenance. Electrical and mechanical problems that cause excessive heat can be spotted quickly.

Markers in the software identified a 35-degree difference in temperature between the leads in this case. During the next outage, the leads were unwrapped, and a loose crimp connection was discovered (Figure 3.28).

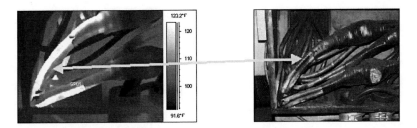

Figure 3.28 Thermographic image of generator leads.

LUBRICATION TRENDING

Machinery that relies on lubrication oils for operation of bearings, gears, and so on requires periodic oil sampling and analysis to find the earliest onset of potential problems. Oil can get contaminated at any time, but more importantly, it will eventually break down and no longer provide adequate protection. Another major benefit of this testing is uncovering the very early breakdown of metals in the machinery. Oil analysis can identify this problem sooner than other tests. Gear face wear, babbitt metal from bearing wear, and so on are all problems that will gradually be identified by vibration analysis, but it will be too late to extend the system's life by purging the system and providing new oil. At the same time, all technologies should be used to understand why the wear is occurring and to ensure maximum machinery life (Figure 3.29).

Oil lubrication requires attention to the height of the oil inside the bearing cavity. Bearing starvation or over-lubrication will result from the improper installation of the oiler. Many sleeve bearings use bronze rings to carry oil from the sump up onto the top of the journal. These rings should be checked for out-of-round conditions and nicks. They should be clean without sharp edges on the inside riding surface. Sharp corners can cause friction and slow the delivery of oil to the journal (see Figure 3.30).

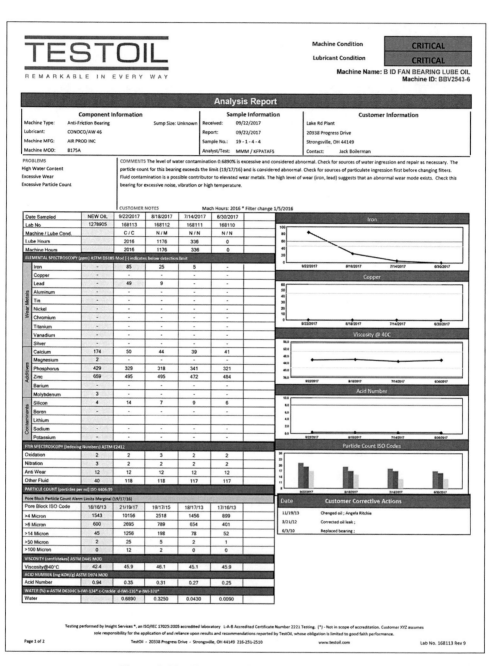

Figure 3.29 Example of oil analysis test results.
(Courtesy of TestOil.)

Figure 3.30   Shop manager James D'Angelo placing an oil reservoir site glass
on a sleeve-bearing feeder.

Another common issue with oil-lubricated bearings is ingestion of oil into the motor windings. Seals can appear to be in good condition and yet the problem may persist. The technician should investigate all passageways inside the bearing and take monometer readings to isolate pressure variations. Ingestion of oil into the motor is most commonly caused by pressure differences. Sometimes it can be as simple as a bearing gasket mistakenly covering an equalization port in the bearing mating surfaces.

Note: T-shaped oil rings should never be installed on equipment operating above 3,000 rpm. A trapezoidal ring is best. There should be zero flat spots on the inside diameter (ID), and the corners of the ID vertical face must be chamfered on a slight angle to prevent the side of the ring from riding on the shoulder of the bearing, causing friction and a slowing of the ring speed.

GREASE LUBRICATION

Companies such as TestOil also can test greases and identify foreign-object contamination. Greases are *not* all the same. Care must be taken for the following:

- **Always** use the properly specified grease for any given machine.
- **Never** combine different greases without full knowledge of their compatibility.

- **Never** over-grease a bearing. (If a little grease is good, is more grease better? *Never!*)
- Do *not* believe that an over-greased bearing will self-purge.

It is my experience that grease never exits the purge port designed to relieve this condition.

Newly installed bearings indicating excess heat are usually over-greased. Don't prolong the agony. Open the bearing and remove some of the grease. A good rule of thumb is to pack a new antifriction bearing with grease on both sides. Install the bearing inside the pillow block housing. Then fill the internal cavity with grease up to the middle of the ball/roller at the bottom of the bearing.

CASE STUDY

A major oil refinery asked my company to look at a 400-hp motor that was running hot. The shop mechanics removed the end bracket, and inside they found a beautiful relief sculpture in lovely blue-green grease. The entire end bracket was full. This wasn't a case of accidental over-lubrication. Someone had hooked up a motorized greaser to this motor and then went to lunch. Easy fix! I had the refinery remove all grease guns and machines from the mechanics and installed automatic lube systems on all motors.

For those of you not familiar with single-point automatic lube systems, they will eliminate all the over- and under-lubrication issues your machinery might encounter. I highly recommend them for one reason: the time period for regreasing bearings might be 6 months. This means that 5½ months down the road, the bearing is most probably under-lubricated. The end user can program this system to provide small amounts of grease each week, guaranteeing that new grease is supplied to the bearing (Figure 3.31).

The L10 bearing life introduced by Lundberg and Palmgren specifies the distance (or number of revolutions, for radial bearings) that 90% of identical bearings, operating under identical conditions, can travel before fatigue (spalling or flaking) occurs on the rolling elements or raceways. (1947 by Gustaf Lundberg and Arvid Palmgren, based on work done at SKF.)

Figure 3.31 Single-point automatic lube systems in operation.
(Courtesy of Automatic Lube Systems, Inc.)

ELECTRICAL TESTING

Motor and generator electrical testing is necessary to trend and then schedule reconditioning. Basic periodic care should include readings with a megohmmeter to verify
that the electrical insulation system is not degrading. A micro-ohmmeter is used to
check for balanced resistance on all windings. A change in winding resistances can
indicate turn-to-turn shorts in a given winding. Depending on where these readings
are taken, they can also uncover issues within cable runs from motor starters to the
motors themselves. Obviously, a problem uncovered between the starter and motor
then must be narrowed down by disconnecting the motor at the motor terminal box
and testing just the motor.

The megohmmeter is also used for the polarization index (PI). This involves
charging the winding and then taking readings during the discharge time. The
10-minute-to-1-minute ratio is calculated and displayed. Readings of 2.0 or higher
are considered acceptable.

Figure 3.32 illustrates an ideal PI test. The motor windings were charged using
500 V. The decay rate should be a parabolic curve. The PI ratio is 2.91. Excellent!

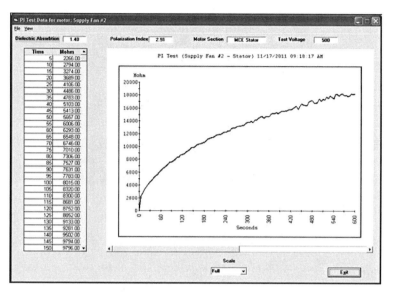

Figure 3.32   Polarization index data with graph.

Electrical equipment testers have become very sophisticated, combining all the above-mentioned tests in a computer-driven automated piece of test equipment. Such equipment not only yields the latest results but also compares against previous test data for any troublesome trends.

Other electrical testing is normally performed during shop reconditioning. These tests include surge testing the windings and a direct-current (DC) hipot test. Alternating-current (AC) hipot testing is required for newly installed winding systems but is *not* recommended for in-service windings. AC hipot testing performed on in-service windings can cause a catastrophic failure of the winding. A shorter version of the PI test is the dielectric absorption (DA) test. A DA test is a 3-minute-to-1-minute ratio of the discharge values. This test was performed on a reconditioned motor winding. Included in the results were peak surge, winding resistance, dc hipot, and the dielectric absorption (Figure 3.33).

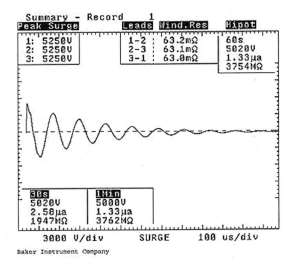

Figure 3.33 Electrical testing results for in-shop reconditioned winding.

Online electrical testing is covered in Chapter 9, dedicated to diagnosing electrical machinery problems.

REVIEW QUESTIONS

1. What is the most important item that should be trended on an oil-filled gearbox?

2. What is the shape of an ideal polarization index test curve on an electrical motor winding?

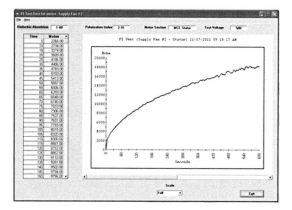

3. What is the significance of the broadband energy at the top center of this historical spectral waterfall plot? Assume ball bearings.

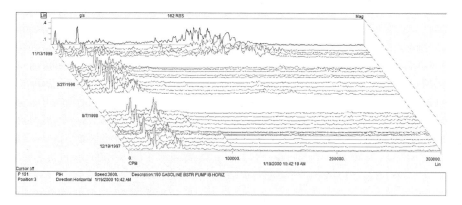

4. What are the two most important software features that should be implemented for a successful vibration trending program?

5. True or False. Gathering data is all that is required for a successful trending program.

6. One alarm band below is raised at vane pass frequency for a pump impeller. How is vane pass frequency calculated?

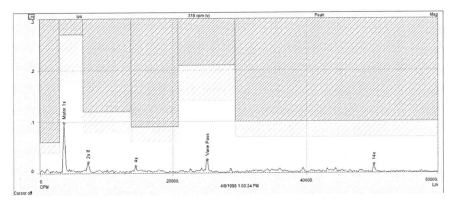

7. A monitored machine has been trending upward for months. Suddenly, the overall vibration and the 1× vibration trends start dropping. What should be the call?

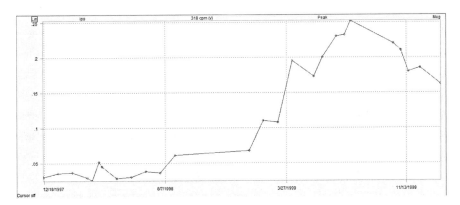

8. True or False. It is best to provide a customer with every piece of recorded data for a final report.

9. True or False. The analyst should provide at least one plot from their instrumentation in the report that substantiates their findings.

10. True or False. There is no need for trending of bearings. Bearings have a calculatable performance life, known as the L_{10} life, and the decision to replace them can be automatically timed by a calendar.

11. True or False. A technician should always be accompanied by an electrician from the customer's site to open and test electrical equipment and never open "live" electrical cabinets.

12. True or False. Thermography can provide an analysis of the condition of load carrying connections in motors and Motor Control Cabinets.

13. Bearings that are oil lubricated can be stationary sumps or can be fed by oil pumps. Where should the oil level be in a bearing with a slinger ring?

14. True or False. A bearing is grease lubricated. Since grease is good for the bearing, more grease is always better.

15. What is the most important aspect of using Automatic Lubers for grease application to a bearing?

16. Electrical testing of motors is a very important Predictive Maintenance procedure. However, cutting away the insulating tape inside the motor terminal box for testing and then re-taping the connections is highly time-consuming and wasteful. Where should initial motor electrical testing be performed?

17. A PI (Polarization Index) Test of a motor winding is performed and the 10-minute ratio is 3.0. The curve is a perfect parabolic shape. What is the verdict?

18. True or False. AC HI pot testing of motor winding systems provides the highest stress available for the evaluation of the winding insulation system. It should therefore be performed ONLY on a new insulation system prior to service.

CHAPTER 4

Machinery Fault Analysis, Part 1

Dynamic Balancing

This chapter begins an extensive look at troubleshooting charts, and how to use them to isolate machinery vibration issues. Dynamic balancing of rotating equipment is a great place to begin, because no matter what the application, there will *always* be a need to assure vibration levels are minimized.

There are charts that explain how various machine faults appear in vibration spectra, how phase is affected, and so on. These charts have been around for years. I penned the example in Figure 4.1 along with fellow vibe technician Peter Phillips in the very early days of our work with narrow-band analyzers (1978). It represents the first time narrow-band fast Fourier transform (FFT) analysis was included in a troubleshooting chart (Envelope Characteristics). Jim Berry of Technical Associates of Charlotte has amassed a very large, detailed chart, and he credits the original Nicolet Scientific chart (later Wavetek).

A more up-to-date machinery troubleshooting guide is provided in Appendix A.

The end goal of writing this book is to assist anyone working in the field to amass their own wealth of successful case histories. Each job is a learning experience, and that knowledge will add up over time until the day you can look at a spectrum someone else has recorded and tell them what the problem is.

NICOLET SCIENTIFIC CORPORATION

MACHINERY VIBRATION DIAGNOSTIC GUIDE

245 Livingston St., Northvale, N.J. 07647 USA Tel: (201) 767-7100

PROBABLE SOURCE	DISTURBING FREQUENCY	DOMINANT PLANE	PHASE ANGLE RELATIONSHIP	AMPLITUDE	ENVELOPE CHARACTERISTICS	COMMENTS
UNBALANCE						
(1) Mass Imbalance	1x rotor speed	radial. (Axial is higher on overhung rotors)	1. force-in-phase 2. couple - 90° out-of-phase or greater	steady	narrowband	Rotor bow due to thermal stresses - may cause change in amplitude and phase with time
(2) Bent Shaft	1x rpm, (2x rpm if bend at the coupling)	axial	180° out-of-phase axially	steady	narrowband	Run-out at rotor mass appears as unbalance. Run-out at coupling appears as misalignment.
(3) Eccentric Motor Rotor	1x rpm, 1x and 2x line frequency	radial	N/A	steady (see comment)	narrowband	Will fluctuate (beat) in amplitude if electrical problem also exists. (see Electrically Induced)
MISALIGNMENT						
(1) Parallel	1x, 2x rpm	radial	180° out-of-phase radials			Most misalignments will be a combination. Errors are most common in the vertical plane. On long coupling spans, 1x will be higher.
(2) Angular	1x, 2x rpm	axial	180° out-of-phase axials	steady	narrowband	
(3) Both parallel and angular	1x, 2x rpm	radial and axial	180° out-of-phase radial and axial			
ELECTRICALLY INDUCED All electrically caused problems can be isolated, i.e. eliminated by cutting the current to motor.						Will fluctuate in amplitude if mechanical problem (unbalance) also exists.
(1) Eccentric Rotor	1x rpm, 1x and 2x line frequency	radial	N/A	steady		
(2) Loose Stator Laminations	high frequency (over 60K cpm)	radial	N/A	high, steady	narrowband sidebands at 2x line F	Not usually destructive
(3) Broken Rotor Bar	running speed with 2x slip frequency sidebands	radial	N/A	steady	narrow spike with sidebands	Only occurs on induction motors. Slip frequency = line frequency minus rotational speed.
(4) Unbalanced Line Voltage	2x line F	radial	N/A	low, steady	narrowband	
(5) Stator Problems (heating, shorts)	2x line F with 2x slip frequency sidebands	radial (can cause axial)	N/A	steady	narrowband with sidebands	
(6) Loose Iron	2x line F	radial	N/A	high, steady	narrowband	
DEFECTIVE BEARINGS						
(1) Anti-Friction	Early stages - 30K-60K cpm depending on size and speed. Late stages - high 1x and multiple harmonics	radial, except higher axial on thrust bearing	N/A	increases as bearing degrades; may disappear just before failure	shows broadband as bearing degrades. Baseline may increase to 40 dB across entire spectrum	Be careful not to mistake for pump cavitation. Baseline signatures will be helpful
(2) Sleeve	Early stages - sub-harmonics (may only be noticeable on shaft) Late stages - will appear as mechanical looseness (see below)	radial	Shaft proximity probe orbits will indicate shaft position and dynamic changes	increase as bearing degrades	high baseline energies below 1x, 2x and 3x rpm	Monitoring of rotor position (thrust) via proximity probes can provide reliable protection against thrust bearing failure
MECHANICAL LOOSENESS						
(1) Bearings, Pedestals, etc. (non-rotating)	1x, 2x, and 3x predominant may be up to 10x at lower amplitude	radial	varies with type of looseness	steady	may extend up to 10x shaft speed	
(2) Impellers, etc. (rotating)	1x predominant, but may have harmonics up to 10x at low levels	radial	will vary from start-up to start-up	steady while running, but will vary from start-up to start-up	may extend up to 10x shaft speed	Variance of amplitude or phase may be caused by center of gravity shifts
OPERATION (process-related)						
(1) Blade/Vane Pass	no. of blades/vane x rpm	radial predominant in the direction of discharge piping	N/A	fluctuating	broadband	More than one (1) discharge volute will produce harmonics of blade passing frequency
(2) Cavitation or Starvation	random - broadband	N/A	N/A	fluctuating	broadband up to 2000+ Hz	
DRIVE BELTS						
(1) Mis-matched, worn or stretched - also applies to adjustable sheave applications	many multiples of belt frequency - but 2x belt frequency usually dominant	radial, especially high in line with belts	N/A	may be unsteady and beating if a belt freq. is close to driver or driven speed	N/A	Be careful not to mistake for pump cavitation. Baseline signatures will be helpful
(2) Eccentric and/or Unbalanced Sheaves	1x shaft speed	radial	in-phase	steady	N/A	Balancing possible with washers applied to taper lock bolts.
(3) Drive belt or sheave face misalignment	1x driver shaft	axial	in-phase	steady	N/A	Confirm with strobe light and belt excitation techniques. Change belt tension or belt length to eliminate the problem.
(4) Drive belt resonance	belt resonance with no relationship to rotational speeds	radial	N/A	may be unsteady	± 20 percent of resonant frequency depending upon damping	
RESONANCE	requires forcing function to excite its natural frequencies	axial or radial	A center hung rotor resonance will display 180° out-of-phase bearing relationships. A component within a structure will display phase relationships dependent upon the bending mode excited.	steady, but baseline energy fluctuations depends on force and damping	appears broad at base, width depends upon damping	Frequency is independent of speed changes
INSTABILITY						
(1) Oil Whirl	40 to 45 percent of running speed with harmonics	radial	N/A	steady 20 to 35 percent of 1x rpm (or higher than 1x rpm in severe cases)	discreet peaks	May excite rotor critical
(2) Rotor Rub	50 percent of running speed and half harmonics	radial	N/A	steady 30 to 35 percent of 1x rpm or higher than 1x rpm in severe cases	discreet peaks	May excite rotor critical
GEARS						
(1) Transmission Error (poorly finished tooth face)	gear mesh frequency (gear rpm x no. of teeth) and harmonics	radial for spur gears, axial for helical or herringbone gears	N/A	depends on loading, speed and total transmission error	usually single peak, but sometimes low sidebands	May excite lateral or torsional resonances at various frequencies. Machining errors during hobbing can cause high 2x or 3x gear rpm vibration.
(2) High Line Runout, Mass Unbalance, Misalignment or Faulty Tooth	1x rpm and gear mesh frequency with 2 gear rpm sidebands	radial for spur gears, axial for helical or herringbone gears	N/A	1x rpm and mesh frequency sidebands depend on fault severity	discreet peaks	

NOTE 1: This guide is based on casing measurements unless otherwise noted.
NOTE 2: Vibration symptoms are based upon the observation of amplitude on a logarithmic scale.

Figure 4.1 Original FFT-based machinery vibration diagnostic guide, 1978.
(Source: Nicolet Scientific Corporation.)

CASE STUDY

I was asked to analyze three vertical pump systems in a new building. The analysis consisted of taking vibration readings on the motors (top floor), two intermediate bearings (each one floor below), and the pumps (basement). The vibration data indicated that the motors, bearings, and pumps were out of alignment, which was serious because they were mounted to weldments on steel beams or sitting on concrete pedestals (pumps). The contractor was adamant that this was not the problem and wanted to hire his own vibration analyst. Everyone agreed.

Two weeks later, I was called back for a meeting at the site. The other analyst began to give his explanation as to why he believed the machine alignment was good. In the middle of his first sentence, he explained his reasoning according to a troubleshooting chart—yes, the chart authored by Peter Phillips and me. Shortest meeting on record!

I would be remiss if I did not mention another troubleshooting chart by John Sohre. This chart (Figure 4.2) was specific to diagnosing problems with turbo machinery. The chart was only legible if printed on legal-sized paper, and it still filled four sheets!

One thing about all the troubleshooting charts. They are a starting point. The technician or engineer should never consult the chart with the data and then make a solid conclusion based on that chart review. Testing *must* be performed to conclusively identify the problem or eliminate the problem through corrective measures. If time on job site won't allow for the implementation of corrective measures, they should be included in the final report to the customer. Corrective measures ought to be the most cost-effective solution to get the job done properly.

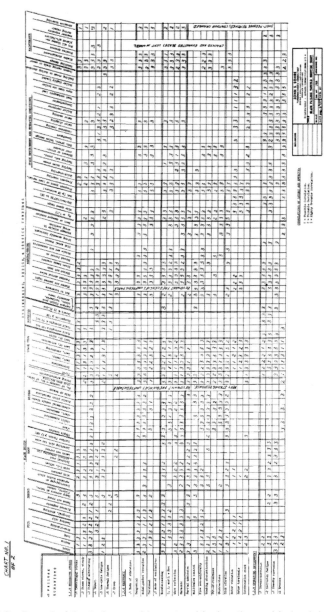

Figure 4.2 Page 1 of the turbo machinery troubleshooting guide by John Sohre. (Available at https://oaktrust.library.tamu.edu.)

DISSECTING A NARROW-BAND FFT
IS CRITICAL TO ANY ANALYSIS

Many spectra have numerous peaks that make it difficult to see the primary causes (Figure 4.3).

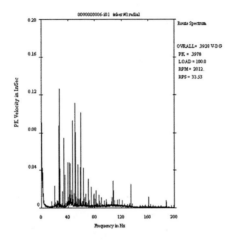

Figure 4.3   Complex FFT spectrum.

The same data with harmonic cursors illustrates that most of the vibration energy consists of harmonics of running speed. However, the correct fault diagnosis lies in identifying the major peak that is *not* a harmonic of running speed (Figure 4.4).

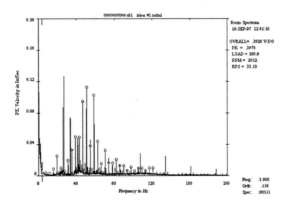

Figure 4.4   Harmonic cursors identify peaks.

CASE STUDY

I was asked to visit a plant in Buenos Aires, Argentina. Their equipment produces machine-painted aluminum beverage cans at rates up to 1,600 cans per minute. These machines have multiple rolls rotating at different speeds. The predominant peak turned out to be another roll that had an imbalance. I traveled out of country to diagnose this problem. Any rotating part should always be checked for balance before being put into operation.

IMBALANCE

One of the most common and easily recognized and corrected causes of vibration is imbalance. What are some of the causes of imbalance? A cast sheave with a blowhole or void in the casting will require that the sheave be balanced. Missing keys or keys not cut to the correct size will cause vibration. Dirt buildup, if uneven, will cause vibration. Usually, the dirt will build up evenly, but then if a large enough piece flies off the rotor, it will be a source of imbalance. Varnish that is sprayed on unevenly or left to pool on one side of a rotor will cause imbalance. Shaft boreholes in rotors that are off-center will cause imbalance vibration. Imbalance also can result from a design issue, for example, operating a rotor at higher than the recommended speed. This happens often when management requests higher production rates without consulting the original equipment manufacturer (OEM) for maximum operating speeds. This will usually result in vibration because the critical frequency of a rotor is encountered, and the rotor bows, causing eccentric rotation and therefore higher than normal 1× vibration.

The imbalance caused by missing keys or keys cut to the wrong length is very common. Since pump manufacturers don't know what couplings their customers will use, they usually provide a full-length key that is shipped with the pump. The installer must cut the key to the correct length once the coupling is procured. It is not uncommon to get to a new construction site and see the full-length key in every single pump installed. The formula for cutting is as follows:

(Keyway length in pump shaft + keyway length in coupling hub)/2 = key length

Once the coupling and key are mounted to the pump shaft, it becomes obvious that the key protrudes from the coupling hub, filling half the remaining keyway in the pump shaft. Thus, if the portion of the key that is visible outside the pump coupling hub is cut off even with the shaft outside diameter, that piece should be big enough to fill the remaining empty keyway. There is usually a curved end in the keyway as a result of the machining. Filling one-half of this portion is usually not critical (Figure 4.5).

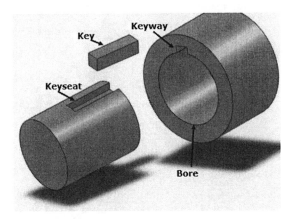

Figure 4.5 Shaft keyway, key stock, and Hub.
(Source: Couplinganswers.com, powered by Lovejoy.)

The main reason these procedures are necessary is because most rotating equipment is factory balanced *before* the keyway is cut into the shaft. Equipment that already has a keyway in the shaft is balanced with a *half-key*, or key stock that fills the keyway even with the shaft outside diameter (OD).

There are times when the correction weight position isn't at a location for attaching the weight. For example when the weight needs to be placed between two fan blades. Technicians can use formulas built into their balance software to split the weight between two existing blades. A graphical representation of the calculation is provided in Figure 4.6.

Steve Young has produced a suite of excellent balance apps for smart phones (Figure 4.7). His apps can be found at www.rotor.zone.

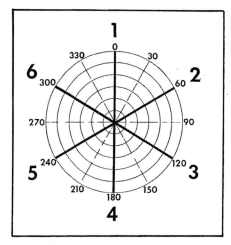

A. On polar graph paper,
mark the relative angular
locations where weight corrections can be made.

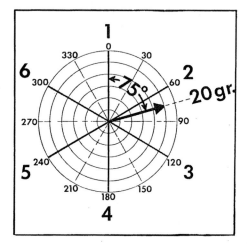

B. Construct a vector repre-
senting the required balance
correction.

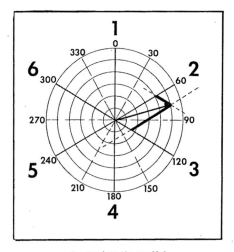

C. Complete the parallel-
ogram.

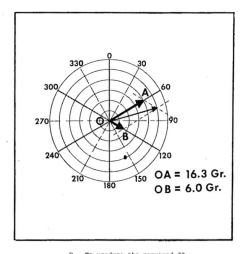

D. To produce the required 20
grams at 75°, 16.3 grams are
needed on blade #2 and 6.0
grams are needed on blade #3.

Figure 4.6   Graphical weight splitting.

(Courtesy of IRD LLC; Application Note 111.)

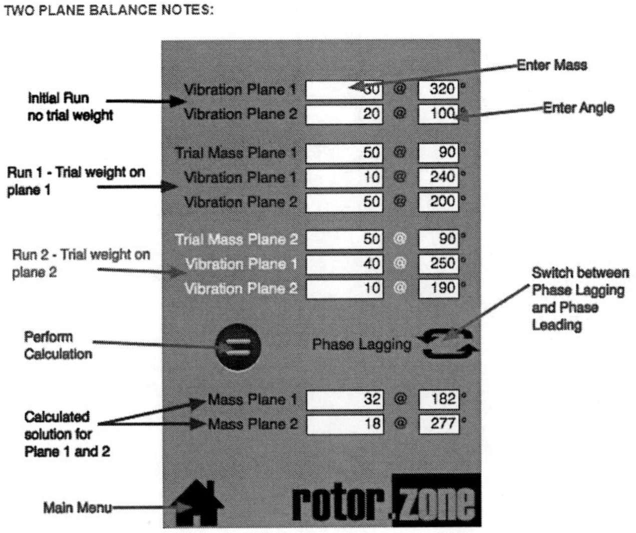

Figure 4.7 Example of a two-plane balancing program for a smart phone. (Courtesy of Steve Young; find his app at: www.rotor.zone.)

Often a balance procedure can result in multiple weights on each end of the rotor. Included with many balance programs is a calculator to combine all weights on one plane into one weight (Figure 4.8).

The *static-couple derivation balance method* will always result in more than one weight on each plane because the method adds weights in pairs to both planes to address the static and couple components of the imbalance. The method examines the original phase relationship of the two bearings, and the user makes the determination of whether the predominant imbalance is static or a couple. Start with the highest reading, and reevaluate after each run.

I watched an IRD field engineer perform this method on a synchronous condenser that had given me fits for days. None of the other balance programs worked. It was later discovered that the unit was operating in close proximity to a shaft-bending mode. The early programs used matrix calculations that did not work well with shafts near resonance. The determinant would approach zero, and as a result, the program would return a correction weight that was excessive. An unknowing technician could follow the program with catastrophic results.

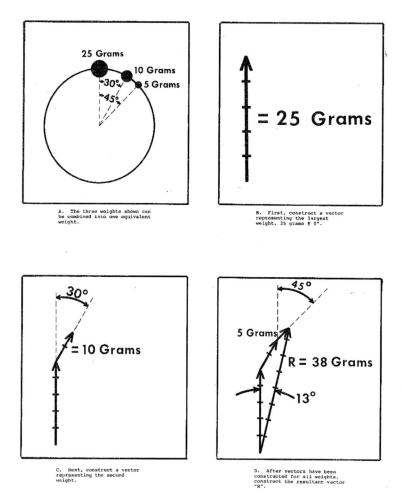

Figure 4.8   Weight consolidation.
(Courtesy of IRD LLC; Application Note 111.)

John M. Csokmay, IRD Mechanalysis, Inc., wrote a paper titled, "Rotor Balancing by Static–Couple Derivation." This paper provided a graphical representation of this balancing procedure. Figure 4.9 is taken from this paper and illustrates the vectors necessary for the procedure.

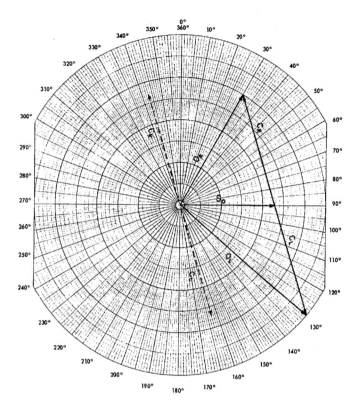

Figure 4.9   Static–couple vectors.
(Courtesy of IRD LLC.)

The static–couple balancing method requires the two original vectors to be drawn first on graph paper (O_L and O_R). The end points of these two vectors are then connected by the resultant vector. The midpoint of the resultant vector is then connected to the origin. This vector represents the *static* portion of the unbalance. The resultant vector is transposed back through the origin. Because couple weights require two simultaneous weights at opposite positions on both planes, either end of the *couple* vector is correct.

The technician then chooses the largest of the imbalance issues and attacks it first. If the static vector is larger than either half of the couple vector, then two weights are positioned on either side of the rotor at the same position. This example calls for the couple imbalance to be addressed first.

Now consider a simple task—balancing a fan in the field. Many customers have fans that require this service. Sometimes the imbalance is due to abrasive wear, erosion, or dirt buildup that, when cleaned, uncovered an imbalanced condition.

CASE STUDY

In the case in Figure 4.10, initial readings indicated that the fan required trim balancing to lower the vibration levels. The data in the figure were recorded before and after the balancing procedure. The customer has a clear indication of the results of the balancing. Fan 1× vibration has been reduced by 75%. Total time for the balance procedure is also captured on the magnitude trace (upper left): 42 minutes—not bad!

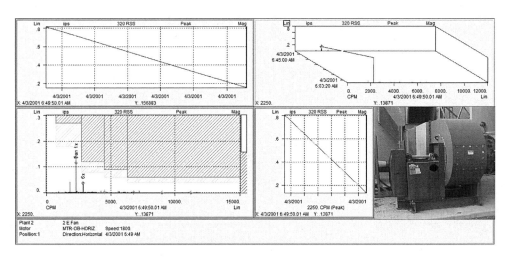

Figure 4.10 Field fan balancing.

This case illustrates the advantage of taking the vibration data using a database template. Both initial and final readings are recorded and plotted, giving the customer the full picture of the completed task. This plot illustrates the hyperlinking feature of Rockwell Emonitor software. The picture of the machine is hyperlinked to the data plot.

The axial vibration on the motor addressed in Figure 4.11 was not reduced by balancing the fan. The motor 1× vibration was excessive because of new belts and a soft motor foundation. New belts tend to vibrate owing to the bend they get from storage. Continued running eventually stretches the bends from the belts.

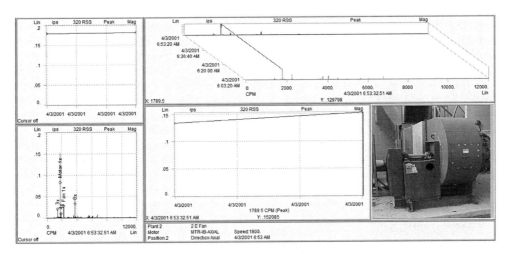

Figure 4.11 Before and after balance vibration at motor bearing axial.

A thorough examination of the rotation assembly always should precede balancing attempts. The fan in Figure 4.12 was scheduled for balancing because of high vibration. An inspection of the fan blade identified a large crack. The blade was repaired on site, and a balance attempt was made. If the imbalance is minor before welding, the welder should pool the same amount of welding rod on the blade 180 degrees from the repair. This will prevent the creation of a large imbalance due to the added weld on one blade.

Figure 4.12 Cracked fan blade.

Caution: When welding cracks or final balance weights to a rotating assembly, *always* ground the welding machine to the fan blade. If the ground is attached to the frame of the machine, arcing can occur between the shaft journal and the bearing components. This type of damage is permanent. The bearings will never yield full life. *Note:* A 1/8" welding rod adds 20 grams of weight.

This fan's vibration in the horizontal direction indicated further issues with this machine (Figure 4.13). A peak averaged plot is beneficial for differentiating between imbalance response and resonance response. The force due to imbalance increases as a square of the speed change (assuming a linear system). The technician can use this formula to identify which of the two is responsible for the running speed vibration. The vibration at full speed (1,800 rpm) is slightly less than 0.5 inch/s. The vibration at 900 rpm (cursor line) was 0.013 inch/s. Doubling the speed from 900 to 1,800 rpm should result in an increase in the 1× vibration of 4×: that is, 0.013 × 4 = 0.052 inch/s.

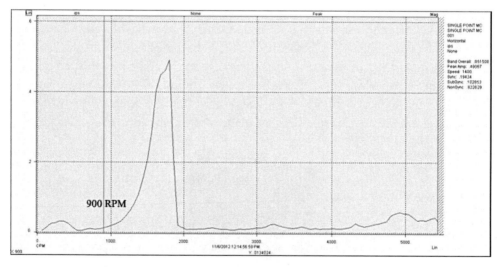

Figure 4.13   Peak averaged plot of fan startup after balancing, horizontal axis.

The readings taken down the entire structure of this fan indicate that the concrete pedestal is rocking, most probably due to voids beneath the foundation (Figure 4.14). The weakness of the foundation has lowered the first rigid-body mode to running speed. This is the cause of the high horizontal 1× vibration on this fan. Trim balancing the fan was successful but was extremely sensitive to small changes of weight and position on the fan blade. A small angular movement of the same trial weight meant the difference between an unsuccessful balance (0.27 inch/s) and a good one (0.11 inch/s). This indicated two separate problems—horizontal resonance and a weak concrete foundation. The customer (contractor operating the bag house) could not convince his customer (the power plant) to correct the foundation issues.

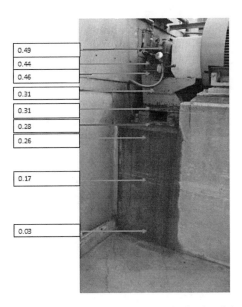

Figure 4.14   Horizontal vibration along the vertical axis of the fan pedestal.

Large vertical motors can be balanced before shipment to acceptable levels, yet when placed on the pump head and coupled to the pump, the resulting vibration is unacceptable.

CASE STUDY

Figure 4.15 was recorded on an 800-hp, 897 rpm motor–pump assembly. The motor had been tested in the motor shop before shipment, as well as in the customer's warehouse, before being moved to a pump house 200 ft from shore. The resulting vibration was very troublesome for the customer because a tug with a crane had been required to move the motor to the pump house.

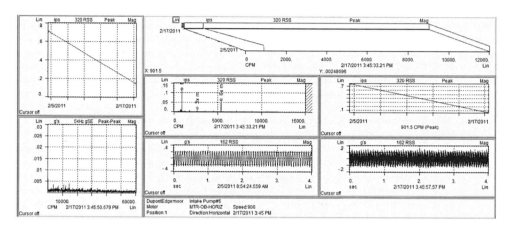

Figure 4.15 Results of balancing a vertical motor.

A resonant condition was suspected, and a simple trim balance was performed on the top anti-rotation ring under the motor top cover. Approximately 40 g of washers was added to one of the three anti-rotation ring connection bolts. This reduced the vibration from 0.7 to 0.16 inch/s. Notice that there is still a peak at 5,344 cpm. This is the vane pass frequency of the pump. A restriction in the discharge piping within the factory was causing an excessive head pressure in the pump (Figure 4.16).

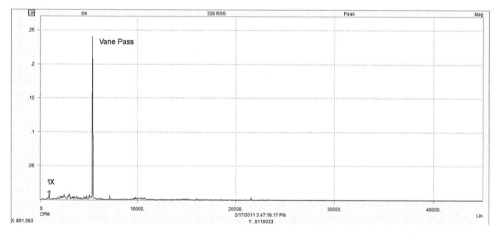

Figure 4.16 Motor drive end axial vibration at vane pass due to excessive head pressure.

This example clearly illustrates that when one mechanical issue is addressed, the result may indicate another totally separate problem. Nothing involving process problems (insufficient suction head pressure, excessive discharge pressure head, etc.) can be solved by balancing the rotating assembly.

Figure 4.17 is a single-plane balancing matrix that was part of a technical article written by William B. Fagerstrom titled, "Balance Calculations Using a Pocket Programmable Calculator." Bill wrote a program for two-plane dynamic balancing using a Texas Instruments TI 59 calculator with a total of 960 bytes of memory. It also included trial weight estimation, weight splitting, and weight consolidation as well.

TABLE 1 — SINGLE PLANE VECTOR METHOD

V_2/V_1	$\Phi_1-\Phi_2=0°$ Ang	Factor	$\Phi_1-\Phi_2=15°$ Ang	Factor	$\Phi_1-\Phi_2=30°$ Ang	Factor	$\Phi_1-\Phi_2=45°$ Ang	Factor	$\Phi_1-\Phi_2=60°$ Ang	Factor	$\Phi_1-\Phi_2=75°$ Ang	Factor	$\Phi_1-\Phi_2=90°$ Ang	Factor	$\Phi_1-\Phi_2=105°$ Ang	Factor	$\Phi_1-\Phi_2=120°$ Ang	Factor	$\Phi_1-\Phi_2=135°$ Ang	Factor	$\Phi_1-\Phi_2=150°$ Ang	Factor	$\Phi_1-\Phi_2=165°$ Ang	Factor	$\Phi_1-\Phi_2=180°$ Ang	Factor
0.05	0	1.05	0	1.05	1	1.04	2	1.03	2	1.02	2	1.01	2	0.99	2	0.98	2	0.97	2	0.97	1	0.96	1	0.95	0	0.95
0.10	0	1.11	1	1.10	3	1.09	4	1.07	5	1.04	5	1.02	5	0.99	5	0.97	4	0.94	3	0.93	2	0.91	1	0.91	0	0.90
0.15	0	1.17	2	1.16	4	1.14	6	1.11	7	1.07	8	1.02	8	0.98	7	0.95	6	0.92	5	0.90	3	0.88	1	0.87	0	0.86
0.20	0	1.26	3	1.23	6	1.20	9	1.14	10	1.09	11	1.02	11	0.98	10	0.93	8	0.89	7	0.86	5	0.84	2	0.83	0	0.83
0.25	0	1.33	4	1.31	8	1.26	12	1.18	13	1.10	14	1.03	14	0.97	12	0.91	10	0.87	8	0.84	5	0.81	2	0.80	0	0.80
0.30	0	1.42	6	1.39	11	1.32	15	1.22	16	1.12	17	1.03	16	0.95	15	0.89	12	0.84	9	0.81	6	0.78	3	0.77	0	0.76
0.35	0	1.53	7	1.49	14	1.39	18	1.26	20	1.13	20	1.03	19	0.94	17	0.87	14	0.82	11	0.78	7	0.76	3	0.74	0	0.74
0.40	0	1.66	9	1.60	17	1.46	21	1.29	23	1.14	23	1.02	21	0.92	19	0.85	16	0.80	12	0.76	8	0.73	4	0.71	0	0.71
0.45	0	1.81	11	1.73	20	1.53	25	1.32	26	1.15	26	1.01	24	0.91	21	0.83	17	0.77	13	0.73	9	0.71	4	0.69	0	0.68
0.50	0	2.00	14	1.87	23	1.61	28	1.35	30	1.15	29	1.00	26	0.89	23	0.81	19	0.75	14	0.71	9	0.68	4	0.67	0	0.68
0.55	0	2.22	16	2.04	27	1.69	32	1.38	33	1.15	31	0.99	28	0.87	24	0.79	20	0.73	15	0.69	10	0.66	5	0.65	0	0.64
0.60	0	2.50	20	2.23	31	1.76	36	1.39	34	1.14	34	0.97	30	0.85	26	0.77	21	0.71	16	0.67	11	0.64	5	0.63	0	0.62
0.65	0	2.85	24	2.44	36	1.83	40	1.40	39	1.13	37	0.95	33	0.83	28	0.75	23	0.69	17	0.65	11	0.62	5	0.61	0	0.60
0.70	0	3.33	29	2.69	41	1.89	44	1.41	43	1.12	39	0.94	34	0.81	29	0.73	24	0.67	18	0.63	12	0.60	6	0.58	0	0.58
0.75	0	4.00	35	2.96	48	1.94	48	1.41	46	1.10	41	0.92	36	0.80	31	0.71	25	0.65	19	0.61	12	0.59	6	0.57	0	0.57
0.80	*	*	42	3.25	52	1.98	52	1.40	49	1.09	44	0.90	38	0.78	32	0.69	26	0.64	19	0.60	13	0.57	6	0.56	0	0.55
0.85	*	*	50	3.52	58	1.99	56	1.38	52	1.07	46	0.88	40	0.76	33	0.68	27	0.62	20	0.58	13	0.55	6	0.54	0	0.54
0.90	*	*	*	*	63	1.99	60	1.36	54	1.04	48	0.86	41	0.74	35	0.66	28	0.60	21	0.56	14	0.54	7	0.53	0	0.52
0.95	*	*	*	*	69	1.97	63	1.33	57	1.02	50	0.84	43	0.72	36	0.64	29	0.59	21	0.55	14	0.53	7	0.51	0	0.51
1.0	*	*	*	*	75	1.93	67	1.30	60	1.00	52	0.82	45	0.70	37	0.63	30	0.57	22	0.54	15	0.51	7	0.50	0	0.50
1.1	*	*	*	*	85	1.81	74	1.23	64	0.94	56	0.78	47	0.67	39	0.59	31	0.54	23	0.51	15	0.49	7	0.48	0	0.47
1.2	*	*	*	*	93	1.66	79	1.16	68	0.89	59	0.74	50	0.64	41	0.57	33	0.52	24	0.49	16	0.47	8	0.45	0	0.45
1.3	*	*	*	*	100	1.51	84	1.08	72	0.84	62	0.70	52	0.60	43	0.54	34	0.50	25	0.46	17	0.44	8	0.43	0	0.43
1.4	*	*	134	1.97	106	1.36	89	1.01	76	0.80	64	0.66	54	0.58	44	0.52	35	0.47	26	0.44	17	0.43	8	0.42	0	0.41
1.5	*	*	139	1.68	111	1.23	93	0.94	79	0.75	67	0.63	56	0.55	46	0.49	36	0.45	27	0.43	18	0.41	9	0.40	0	0.40
1.6	180	1.66	142	1.46	115	1.12	96	0.87	81	0.71	69	0.60	58	0.53	47	0.47	37	0.44	27	0.41	18	0.39	9	0.38	0	0.38
1.7	180	1.42	145	1.28	119	1.02	99	0.82	84	0.67	71	0.57	59	0.50	48	0.45	38	0.42	28	0.39	18	0.38	9	0.37	0	0.37
1.8	180	1.25	147	1.14	121	0.94	102	0.78	86	0.64	72	0.54	60	0.48	49	0.43	39	0.40	29	0.38	19	0.36	9	0.35	0	0.35
1.9	180	1.11	149	1.03	124	0.87	104	0.72	88	0.60	74	0.52	62	0.46	50	0.42	40	0.39	29	0.37	19	0.35	9	0.34	0	0.34
2.0	180	1.00	150	0.93	126	0.80	106	0.67	90	0.57	75	0.50	63	0.44	51	0.40	40	0.37	30	0.35	20	0.34	10	0.33	0	0.33
2.2	180	0.83	153	0.79	129	0.70	109	0.60	93	0.52	78	0.46	65	0.41	53	0.37	42	0.35	31	0.33	20	0.32	10	0.31	0	0.31
2.4	180	0.71	154	0.66	131	0.61	112	0.54	95	0.47	80	0.42	67	0.38	55	0.35	43	0.33	32	0.31	21	0.30	10	0.29	0	0.29
2.6	180	0.62	156	0.59	133	0.55	114	0.49	97	0.44	82	0.39	66	0.36	56	0.33	44	0.31	32	0.29	21	0.28	10	0.27	0	0.27
2.8	180	0.55	156	0.53	135	0.50	116	0.45	99	0.40	84	0.36	70	0.33	57	0.31	45	0.29	33	0.27	22	0.27	11	0.26	0	0.26
3.0	180	0.50	157	0.48	136	0.45	117	0.41	100	0.37	85	0.34	71	0.31	58	0.29	46	0.27	34	0.26	22	0.25	11	0.25	0	0.25
3.2	180	0.45	158	0.44	137	0.41	119	0.38	102	0.35	86	0.32	72	0.29	59	0.27	46	0.26	34	0.25	23	0.24	11	0.23	0	0.23
3.4	180	0.41	158	0.40	138	0.38	120	0.35	103	0.33	87	0.30	73	0.28	60	0.26	47	0.25	35	0.23	23	0.23	11	0.22	0	0.22
3.6	180	0.38	159	0.37	139	0.35	121	0.33	104	0.31	88	0.28	74	0.26	60	0.25	48	0.23	35	0.22	23	0.22	11	0.21	0	0.21
3.8	180	0.35	159	0.35	140	0.33	122	0.31	105	0.29	89	0.27	75	0.25	61	0.23	48	0.22	36	0.21	23	0.21	11	0.20	0	0.20
4.0	180	0.33	160	0.33	140	0.31	122	0.29	106	0.27	90	0.25	75	0.24	62	0.22	49	0.21	36	0.21	24	0.20	12	0.20	0	0.20

Figure 4.17 Single-plane balance procedure matrix.

Directions for Using the Matrix

- * = trial weight not heavy enough
- V1 = vibration amplitude with no trial weight
- V2 = vibration amplitude with trial weight
  - Amplitudes can be measured with any vibration units as long as the same units are maintained
- Φ_1 = shaft phase reading with no trial weight
- Φ_2 = shaft phase reading with trial weight
- Ang = angle to move the trial weight;
  - Move weight opposite the phase shift between Φ_1 and Φ_2

The correct balance weight equals the trial weight times the F factor.

The preceding examples represent the easiest balancing procedure—single plane. Balancing a rotor in two planes is necessary when the imbalance on each end of the rotor is not at the same angular location. In other words, the imbalance on one end is at 90 degrees and the other end has the imbalance at 270 degrees. This type of problem requires more sophisticated software that not only looks at the trial results of one end but also considers what that trial has done to the magnitude and phase of the other plane. The math required to perform this task is substantial, as can be seen in Figure 4.18.

VECTOR CALCULATIONS FOR TWO-PLANE BALANCING

STEP #	ROTOR CONDITION & RUN NO. OR CALCULATION PROCEDURE		SYMBOL	ITEM NO.	PHASE ANGLE/ WT. LOCATION	ITEM NO.	VIBRATION AMP./ WT. AMOUNT
I	Original Rotor Unbalance (Run No. 1)		N	1	63°	2	8.6
			F	3	206°	4	6.5
II	Near End Trial Weight		TW$_N$	5*	270°	6*	10 oz.
III	Resultant Unbalance — Near End Trial Wt. (Run No. 2)		N$_2$	7	123°	8	5.9
			F$_2$	9	228°	10	4.5
IV	Far End Trial Weight		TW$_F$	11*	180°	12*	12 oz.
V	Resultant Unbalance — Far End Trial Weight (Run No. 3)		N$_3$	13	36°	14	6.2
			F$_3$	15	162°	16	10.4
VI & VII	$A = N_2 - N$ (N → N$_2$)		A	17*	201°	18*	7.6
	$B = F_2 - F$ (F → F$_2$)		B	19*	124°	20*	7.3
	$\alpha A = F_2 - F$ (F → F$_2$)		αA	21	350°	22	2.9
	$\beta B = N_2 - N$ (N → N$_2$)		βB	23	286°	24	4.2
VIII	25 = 21 − 17	26 = 22 ÷ 18	α	25*	149°	26*	.382
	27 = 23 − 19	28 = 24 ÷ 20	β	27*	162°	28*	.575
	29 = 25 + 1	30 = 26 × 2	αN	29	212°	30	3.28
	31 = 27 + 3	32 = 28 × 4	βF	31	8°	32	3.74
IX X	$C = \beta F - N$ (N → βF)		C	33	268°	34	7.15
	$D = \alpha N - F$ (F → αN)		D	35	20°	36	3.3
XI THRU XIV	37 = 25 + 27	38 = 26 × 28	$\alpha\beta$	37	311°	38	.22 units
	PLOT UNITY VECTOR		U	39	0°	40	1.0 units
	$E = U - \alpha\beta$ ($\alpha\beta$ → U)		E	41*	11°	42*	.87 units
XV	43 = 33 − 41	44 = 34 ÷ 42	θA	43	257°	44	8.21
	45 = 35 − 41	46 = 36 ÷ 42	ϕB	45	9°	46	3.8
	47 = 43 − 17	48 = 44 ÷ 18	θ	47	56°	48	1.08
	49 = 45 − 19	50 = 46 ÷ 20	ϕ	49	245°	50	.52
	51 = 5 − 47	52 = 6 × 48	Cr.Wt.$_N$	51	214°	52	10.8 oz.
	53 = 11 − 49	54 = 12 × 50	Cr.Wt.$_F$	53	295°	54	6.24 oz.
	GRAPHIC CHECK OF SOLUTION						
XVI	55 = 49 + 23	56 = 50 × 24	$\phi\beta$B	55	171°	56	2.18
	57 = 47 + 21	58 = 48 × 22	$\theta\alpha$A	57	46°	58	3.13
	X = θA + $\phi\beta$B		X	59	243°	60	8.6
	Y = ϕB + $\theta\alpha$A		Y	61	26°	62	6.5
XVII XVIII	APPLY BALANCE CORRECTIONS						
XIX	CALCULATE ADDITIONAL CORRECTIONS AS REQUIRED						

Figure 4.18   Two-plane balance results and calculation sheet. (Courtesy of IRD LLC; Application Note 111.)

Balance tolerances are wide ranging depending on the size and speed of the rotor. The application can also have an impact on the tolerance chosen. For example, a printing company introduced a laser printing machine for setting type. The laser was reflected by a rotating hexagonal mirror. The mirror ran in air bearings. The specifications called for "as low as you can get it." In order to achieve the lowest vibration of running speed, the output of the tunable filter analyzer was input into a narrow-band FFT analyzer to get the highest-magnitude resolution. When the peak was finally low enough for the company's satisfaction, the last weight addition was 2 g. The mirror rotated at 300 rpm and weighed over 30 lb (Figure 4.19). Obviously, this job would have fallen into the G 0.4 range.

ROTOR CLASSIFICATION (Balance Quality)	ROTOR DESCRIPTION (Examples of General Types)
G 40	Passenger Car Wheels and Rims
G 16	Automotive Drive Shafts Parts of crushing and agricultural machinery.
G 6.3	Drive shafts with special requirements Rotors of processing machinery Centrifuge bowls; Fans Flywheels, Centrifugal pumps General machinery and machine tool parts Standard electric motor armatures
G 2.5	Gas and steam turbines, Blowers, Turbine rotors, Turbo generators, Machine tool drives, Medium and bigger electric motor armature with special requirements, Armatures of fractional hp motors, Pumps with turbine drive
G 1 Precision Balancing G 0.4 Ultra Precision Balancing	Jet engine and super charger rotors Tape recorder and phonograph drives Grinding machine drives Armatures of fractional hp motors with special requirements Armatures, shafts and sheels of precision grinding machines

Figure 4.19 Balance tolerance by application.
(Courtesy of IRD LLC; Application Note 111.)

How much imbalance can the machine live with? The chart in Figure 4.20 was formulated by engineers from around the world at a meeting in Germany in 1940. End users can find on this chart the recommended imbalance tolerances for their

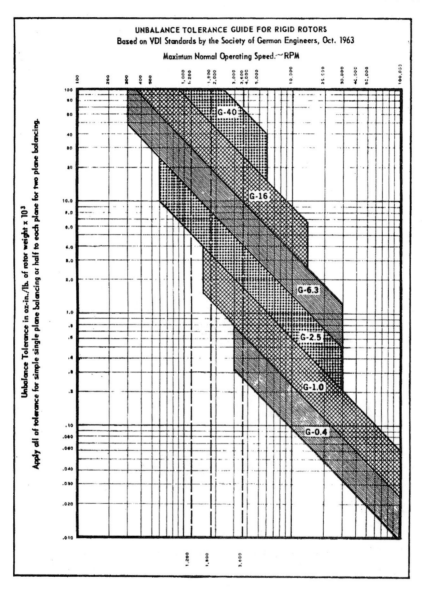

Figure 4.20   Balance tolerances for rigid rotors.
(Courtesy of IRD LLC; Application Note 111.)

machines. Why was a G used for the various levels of tolerance? For years, I mistakenly thought this stood for "grade." It doesn't. It's a German abbreviation. The levels are bracketed by values of *Geschwindigkeit*, which is German for "velocity." A balance tolerance of G 2.5 means that the rotor, if turning free in space at speed, would not vibrate more than 2.5 mm/s. The bands appear the way they do because the tolerance varies based on the speed of the rotor.

The German engineers understood that this limit was a starting point for the final balance and that the operator would strive to do better. The American engineers attending the meeting pleaded with the Germans to change their minds, stating that in America, once the limit is met, the technician or boss would consider the job completed! So please consider these tolerances as the bare minimum to be met, *especially* when balancing two pole rotors. For example, a rotor turning at 3,600 rpm calls for an imbalance tolerance of 2.0 oz/inch per plane per 1,000 lb of rotor weight.

Important: Balance tolerance should never be expressed in vibration units, mils, inch/s, etc. Balance specifications require a weight and a radius. That's it! Then it is irrelevant of speed. For example, a rotor might call for a final balance tolerance of 15 g-inches. This means 15 g at a 1-inch radius. If the correction plane is at a 3-inch radius, then that would call for 5 g. Fortunately for users, the weight-versus-radius changes are linear. Double the radius, half the weight.

There are many examples of balance procedures that do not give the expected results. Often this situation arises from balancing multiple components on a single shaft.

CASE STUDY

An OEM balances multiple components on a single shaft in a precision balancing machine. The components are taken apart and then reassembled in the machine. Vibration levels do not meet specifications. What happened? Multiple components must be match marked after balancing and *before* disassembly. The components must be placed in the same orientation for reassembly. The reason is simple: If the components can slide off the shaft, there is clearance. That clearance, especially with multiple components, can add up to a residual imbalance that will result in excessive vibration.

REVIEW QUESTIONS

1. Troubleshooting charts are an excellent tool and should be used in what context?

2. Corrective measures should be engineered to eliminate the problem but also concentrate on minimizing what?

3. Vibration spectra can have many frequency peaks. What is key to simplifying the analysis of the number of peaks of interest?

4. Unbalance is a fundamental cause of vibration in machinery. What frequency of interest identifies the unbalanced condition of a rotor?

5. A new pump installation is in process. The key fills the entire keyway in the pump shaft. That keyway is 2 inches long. The keyway in the pump coupling is 1 inch long. Does the key need to be cut, and if so, how long should it be? (Ignore the Woodruff curved portion at the end of the pump keyway.)

6. True or False: It's completely acceptable to balance a fan with cracks and/or significant dirt buildup.

7. A trial weight is added to one end of a long rotor. The vibration level decreases on that end but increases on the other end, and the phase stays the same. What caused this result?

8. How much weight is added to a rotor by using two full ⅛-inch-diameter welding rods to attach a final weight?

9. True or False: Welding on a fan blade requires the welding grounding clamp to be placed on the fan impeller itself.

10. Force due to imbalance increases as the square of the speed increases. A 1× vibration at running speed on a fan is 1 inch/s velocity peak. The vibration of the 1× component at 50% of speed is 0.02 inch/s. Is the problem at running speed due to imbalance?

11. Balance of a customer rotor calls for a balance tolerance of G 0.4. What does the G stand for, and what is the significance of the 0.4 level?

12. True or False: A rotor that has been balanced down to the G specification is considered ready to ship.

13. A customer has sent a rotor to your shop for balancing. The balance specification calls for "less than 1 mil vibration." How would you proceed?

14. True or False: A rotating assembly with multiple parts can achieve an acceptable balance by balancing all the components separately and then putting all of them together.

15. Balancing a fan on a motor-driven machine has not dropped the overall vibration sufficiently. What is the next step?

16. Wheels on a passenger train car have a balance tolerance of G 40. Why is this acceptable?

17. A balance run using a trial weight placed at a 6-inch radius has called for a final weight of 210 g. The final weight is to be welded to the face of the fan at a radius of 18 inches using a ⅛-inch welding rod. How much should the steel plate weigh?

18. A fan and shaft are sent to a shop for bearing replacement. The shop balances the fan in a hard bearing machine operating at 450 rpm. The fan is high speed, being directly driven in the field by a motor at 3,600 rpm. The field engineer has cautioned the shop foreman to paint the final balance weight with red paint. Why? (*Note:* Below critical frequency, the fan will rotate around its center of rotation. Above critical frequency, the fan will rotate around its center of mass.)

CHAPTER 5

Machinery Fault Analysis, Part 2

Resonance Effects on 1× Operating Speed Vibration

We have covered how to balance rotating machinery. However, there are other issues that can affect the 1× operating speed response. This chapter illustrates many of those problems and provides the reader with test techniques to still achieve low vibration levels.

THERMALLY SENSITIVE MOTOR ROTORS

The term *thermally sensitive* has been used for years to explain motor rotors that change their balance as they heat up under load. What can change on an induction rotor as it heats up that would change the balance?

The process of assembling a copper or aluminum barred rotor involves heating the assembled rotor in an oven and then dropping it onto the rotor shaft. If this assembly is not kept vertical until the rotor is cold, the rotor core can warp. The rotor is then balanced. Under load, the heat in the core will release the "as manufactured" stress, and the rotor will now have an imbalance.

The solution is to remove the lamination stresses first and then rebalance the rotor cold. This is done by hanging the rotor vertically in an oven at 700°F overnight. The rotor is then removed and beat with dead-blow hammers to destress the laminations. This should not be more than a couple of minutes. The rotor is then covered with a thermal blanket and allowed to cool slowly. The destressed rotor now can be balanced dynamically. This method has resolved issues with numerous motor rotors that had been pulled from service for continuously poor performance.

CASE STUDY

I was asked to visit a field site where a recently reconditioned generator had a thermally sensitive rotor. Vibration levels on startup were acceptable but after 1–2 hours of operation, vibration levels had increased to a shutdown condition. This particular job illustrates the need for the field technician to logically think through what is happening to the rotor.

Phase readings taken on the generator were always in phase not only on startup but even as the levels approached shutdown. This means that balance weights are required to be at the same position on both sides of the rotor. Also, the phase information given by modern data collectors and laser tachometers has little or no phase lag. This means that the technician can easily "guesstimate" the heavy spot and then place the trial correction weights opposite that location.

The vibration levels need to be lowered by the trial weights, but more important, the phase readings need to be 180 degrees opposite the original phase readings. Then, as the thermal bow affects the rotor, the vibration readings will approach very low levels as the phase shifts through zero and approaches the initial phase readings. The startup vibration levels cannot be excessive—they must be tolerable, knowing that they will go down.

This particular generator called for approximately 5 lb of weight on each side of the rotor. Startup vibration levels were at the 0.15- to 0.2-inch/s level. Again, not low, but tolerable. The final vibration readings after a 2-hour full-load run were in the 0.05-inch/s range.

Modern data collectors have single- and two-plane balancing programs built into their software. Understanding what the software does can often explain why a balance procedure isn't working. Figure 5.1 is an example of a single-plane balance procedure. The original imbalance vector was plotted on polar paper. The "O" (original) measurement was 5 mils at 120°C. Remember, balancing requires two pieces of information from every measurement, magnitude, and phase. As far as the magnitude is concerned, the units don't matter, as long as it represents the magnitude of 1× operating speed.

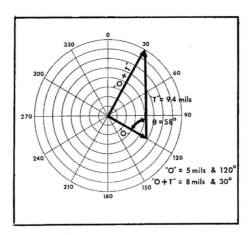

Figure 5.1   Single-plane vectors for static balance.
(Courtesy of IRD LLC; Application Note 111.)

A trial weight then must be selected and placed on the object being balanced. Choosing the right size trial weight is important. Too much weight could cause the machinery to have such high vibration that permanent damage is done. Too little weight might mean that the change in magnitude and phase will not be enough for the calculations to be performed.

The trial weight should be large enough to cause a force equal to 10% of the rotor weight. The following example assumes a rotor weight of 1,000 lb. The formula for choosing a trial weight is as shown in Figure 5.2.

$$\text{Force (lbs)} =$$

$$1.77 \times \left(\frac{\text{RPM}}{1,000}\right)^2 \times \text{ounce-inches}$$

$$100 \text{ lbs} =$$

$$1.77 \times \left(\frac{3,600}{1,000}\right)^2 \times \text{ounce-inches}$$

$$\text{ounce-inches} = 4.36$$

Figure 5.2   Calculation to estimate trial weight.
(Courtesy of IRD LLC; Application Note 111.)

Figure 5.3 is an example of a slip-stick give-away. The user slides the insert to the revolutions per minute (rpm) of the rotor being balanced. Knowing the rotor weight, the top window indicates a suitable trial weight. The original and trial weight vibration runs then occur, and the amplitude and phase angle of each run are recorded. The middle and bottom windows then can be used to calculate the ratio of the vibration of the two runs and the phase-angle shift. Turning the device over, the user slides the left window until the phase shift lines up with the amplitude ratio. The lower window now indicates the amount of correction weight, and the right window shows the angle to move from the trial weight location to where the correction weight should be attached.

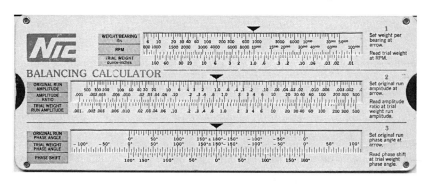

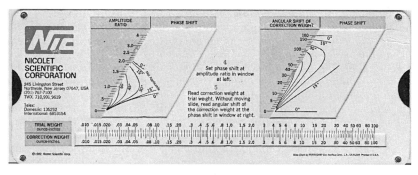

Figure 5.3   Single-plane balance slip stick.
(Courtesy of Nicolet Scientific Corporation.)

Unfortunately, most rotor weights are not known, and this is where experience will help. That said, you are always better off underestimating trial weights as a beginner.

A customer had two fans that were always having balance issues owing to the abrasive product of the system. The fans were 6 ft in diameter and 10 inches wide, rotating at 1,800 rpm, direct drive. A trial weight of 50 g at the edge of the fan (3-ft radius) worked well. Trial weights are best placed on the leading edge of the fan blade. If this is not possible, use steel trial weights and *always* tack weld them on!

Rotors that are operating near a critical frequency will have amplification of the vibration at 1×. This calls for a much smaller trial weight.

CASE STUDY

Another customer had a motor-gearbox-compressor in chlorine service with a compressor speed of 8,400 rpm. There was a natural frequency near this speed. A moderate vibration issue was balanced out successfully with an addition of 10 g to the coupling.

The phase information on the compressor bearings indicated that the ends were 180 degrees out of phase. This would mean that the coupling-rotor combination was operating near the second rigid-body mode. Adding weight to the compressor side of the coupling brought the vibration levels down on both ends of the compressor. If the imbalance had been a couple, meaning that weight would be required on both ends of the rotor placed out of phase with each other, addition of the weight to the coupling would have lowered the vibration on one bearing, but the other bearing levels would have risen.

The matrix in Figure 5.4 can be used to cut trial or final weights from ¼-inch plate steel. An additional "field learned" fact: a ⅛-inch welding rod, when used to attach the steel plate to the rotor, adds approximately 20 g per rod. *Trick:* Hammer all the flux off the rod and then weigh it!

Weight calculations for 1/4 plate steel with various lengths and widths
Density - 0.283 lbs./cubic inch   454 grams = 1 lb.

Length	1	1.25	1.5	1.75	2	2.25	2.5	2.75	3
Width									
1	32.1	40.2	48.2	56.2	64.2	72.3	80.3	88.3	96.4
1.25	40.2	50.2	60.2	70.3	80.3	90.3	100.4	110.4	120.5
1.5	48.2	60.2	72.3	84.3	96.4	108.4	120.5	132.5	144.5
1.75	56.2	70.3	84.3	98.4	112.4	126.5	140.5	154.6	168.6
2	64.2	80.3	96.4	112.4	128.5	144.5	160.6	176.7	192.7
2.25	72.3	90.3	108.4	126.5	144.5	162.6	180.7	198.7	216.8
2.5	80.3	100.4	120.5	140.5	160.6	180.7	200.8	220.8	240.9
2.75	88.3	110.4	132.5	154.6	176.7	198.7	220.8	242.9	265.0
3	96.4	120.5	144.5	168.6	192.7	216.8	240.9	265.0	289.1
3.25	104.4	130.5	156.6	182.7	208.8	234.9	261.0	287.1	313.2
3.5	112.4	140.5	168.6	196.7	224.8	252.9	281.1	309.2	337.3
3.75	120.5	150.6	180.7	210.8	240.9	271.0	301.1	331.2	361.4
4	128.5	160.6	192.7	224.8	257.0	289.1	321.2	353.3	385.4

These answers are in grams

Figure 5.4   Trial weight calculations.

CASE STUDY

A pump impeller had recently been dynamically balanced for a customer and reinstalled in the field. The customer called to complain that vibration levels were very high. A field team was dispatched to the site. The vibration readings were high and all at 1× operating speed. The decision was made to solo the motor to ensure that there were no problems in that part of the machinery. The coupling guard was removed after the motor was locked out. Problem identified! The customer's mechanics had run out of coupling bolts, so they went to the storeroom to get two more bolts. The new bolts were 1 inch longer than the original bolts. The mechanics had placed the longer bolts right next to each other! One of the longer bolts was swapped with a regular bolt that was 180 degrees from the other long bolt. Smooth!

Note: The reader will notice that older case studies are plotted with a log scale on the amplitude axis. I learned early the benefit of viewing diagnostic data in this way. Resonance conditions are much easier to spot, as well as amplifying sideband frequencies for bad bearings, rotor bars, and so on.

CASE STUDY

There are situations where phase information cannot be gathered. A customer had just put new turbine blades in a 30-MW peaking generator in a rural environment. The unit operated above the first bending critical of the rotor shaft. Following a machine overhaul, the turbine would start up, but as the speed approached the first critical, the vibration levels were so high that the unit would trip offline. New blades had been sent from the manufacturer, with blade placement numbers stamped on them to ensure a balanced assembly. It was discovered after the entire machine had been reassembled and sealed with insulation that the blades had been mismarked. This was the cause of the imbalance.

There was no tachometer output on the unit, and the entire rotating assembly was enclosed, so a strobe light or laser tachometer could not be used. There was not even a constant speed because the unit would run up to trip and then coast to a stop. The customer was facing the undesirable inevitability that the turbine would have to be ripped open again to make corrections.

The initial run up of the turbine was captured on a fast Fourier transform (FFT) with peak-averaged spectra. This type of spectrum saves the maximum vibration for each frequency bin for the entire run. Again, this was necessary because the turbine would never reach a constant speed. The proximity probes that were permanently mounted in the gas turbine were used for capture of the magnitude data.

The initial run captured the problem. The vibration level was in excess of 6 mils as it passed 2,700 rpm. The highest vibration captured by the customer was in excess of 8 mils. This was the first critical of the turbine shaft. The turbine tripped, and the data were plotted (Figure 5.5).

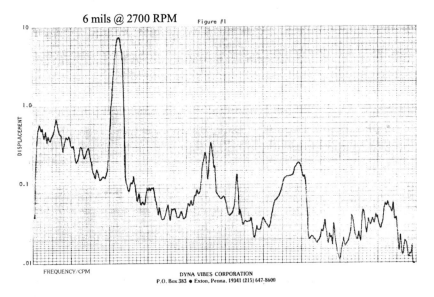

Figure 5.5   Initial vibration peak hold averaged spectrum.

Note: Given the nature of the startup, a speed of 2,700 rpm was chosen to take all magnitude readings because they were captured just before the trip.

Balance weights were provided by the customer. These were hex-head screw plugs weighing approximately 100 g each. The on-site mechanic had to climb down the bore containing the startup drive shaft, remove a plug in the turbine casing, and then screw in the balance plug. The mechanic had to signal the operators when to stop the turning gear (used to prevent shaft sag after shutting down a hot rotor) so that he could insert the plug. The first shot, being a trial, was just a guess. The vibration at 2,700 rpm was 7.7 mils. The bad news: the vibration was worse. The good news: the rotor was responsive to the weight (Figure 5.6).

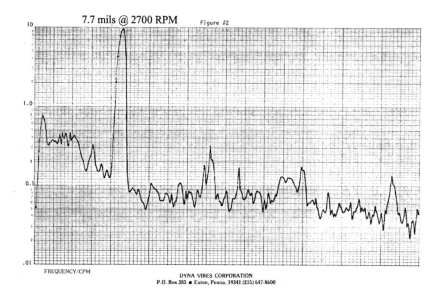

Figure 5.6   First trial weight run.

The trial was removed from the first test position and moved 120 degrees. The resultant vibration with the trial in the next hole was 5.5 mils (Figure 5.7).

The result of the third trial run was 5.4 mils (Figure 5.8). The calculated results called for 3.8× the trial weight. 3× the trial weight was used so as not to overcompensate from operating close to the second critical of the rotor. The final weights were placed equally between the final two trial positions (or 180 degrees from the highest reading at trial position 1).

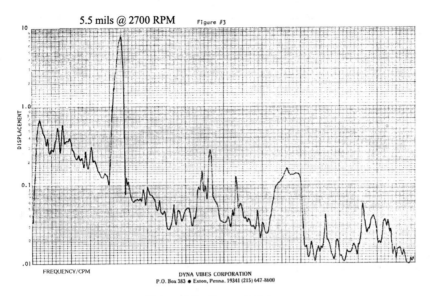

Figure 5.7 Second trial run.

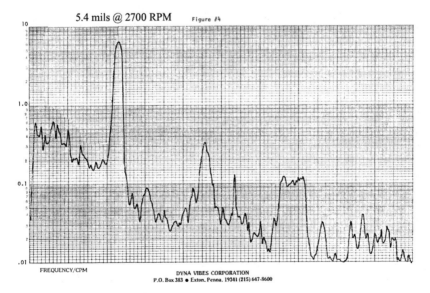

Figure 5.8 Third trial run.

The turbine-generator accelerated and for the first time in a month successfully passed through the first critical and came up to its normal operating speed of 4,985 rpm. The vibration reached 4.5 mils as the rotor passed through the first critical. This was below the trip level (Figure 5.9).

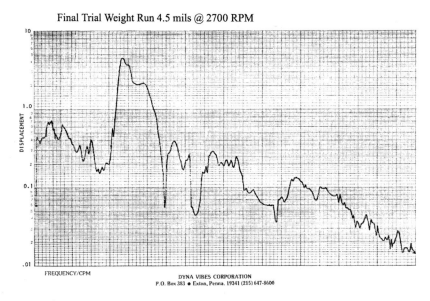

Figure 5.9   Final run with three correction weights at 180 degrees.

The final vibration of the turbine at an operating speed of 4,895 rpm was 0.5 mil (Figure 5.10). Most of the peaks in this spectrum were due to scratches on the target area of the shaft. The overall vibration was less than 1.0 mil on both ends of the rotor. Tolerance specification was 1.5 mils. The customer was ecstatic. This unit hadn't made 1 kW in more than a month. This fix was made without any disassembly of the unit. The customer bought lunch!

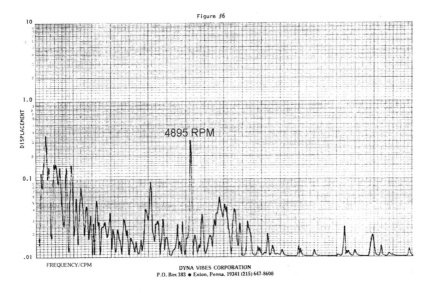

Figure 5.10   Final online vibration.

Earlier in this section, I claimed that two pieces of information were required from each measurement required for balancing a rotor. Well, that's not quite true. The balance procedure for the previous case study is referred to as the *four-run method*. It requires an initial run and then three separate trial weight runs, with the weight moving (120 degrees, ideally) for each run. The technician then takes out his or her pad of paper and trusty compass and proceeds to plot out the data and find the solution.

Steve Young has posted his machinery vibration programs online at www.rotor. zone. These programs include balance with no phase, single- and two-plane balancing, weight splitting, weight consolidation, and so on (Figure 5.11).

The preceding case history concerning the gas turbine with the imbalanced blades tripping on startup was an ideal use for the four-run method without phase information. Below is an example of using the program that is illustrated in Figure 5.11.

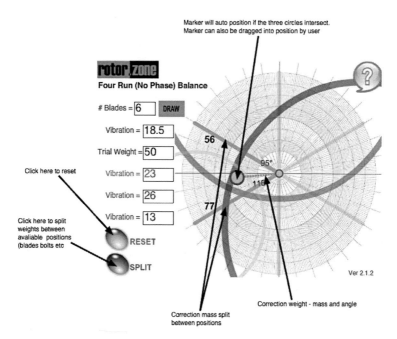

Figure 5.11   Results of four-run, no-phase balancing using app from Steve Young. (Courtesy of Steve Young; App available at: www.rotor.zone.)

The fan in this case had six blades. The initial vibration was 18.5 mils. Fifty-gram weights were placed, one at a time, on three blades separated by 120 degrees. The vibration readings are plotted on the graph in the figure. The intersection of the three results marks the blade where the final correction weight will go. If the intersection is within the original circle, the trial weight was too light. If the intersection is outside the circle, the trial weight was too heavy. In this case, the ratio of the intersection to the original circle is approximately 2.5. The trial weight (50 g) times 2.5 (125 g) is the final correction weight. The final weight also falls between two blades. The app includes the weight-splitting program to find the correct weight to place on the two adjacent blades.

RESONANCE CONDITIONS AND
HOW THEY AFFECT MAGNITUDES

Here is another example of peaks in the FFT that are not harmonics of running speed. These nonsynchronous frequencies are usually related to rolling-element bearings. The predominant peak in the plot below is the first harmonic (2×) of a lower-frequency peak. The peak is 5.145× running speed and was identified as the *ball-pass frequency of the outer race* (BPFO). The predominant peak is the first harmonic of BPFO (2×). The reason this peak has a higher magnitude is a localized natural frequency. The resonant energy is identified by the broad base of the peak. The technician can change the *Y* scaling from linear to logarithmic to enhance the low-amplitude energy such as the base of a resonant peak (Figure 5.12).

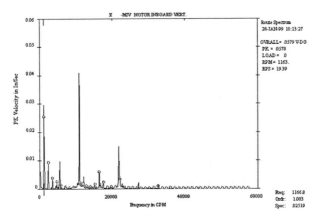

Figure 5.12   Nonsynchronous vibration.

When a localized natural frequency is driven into resonance, the damping associated with that natural frequency can be seen in the broad base under the response at the natural frequency. The plot in Figure 5.13 identifies two such frequencies. The base of the peaks is broad, indicating a significant degree of damping. Using the *Y* log scaling is very effective in amplifying low-magnitude energy such as resonance response.

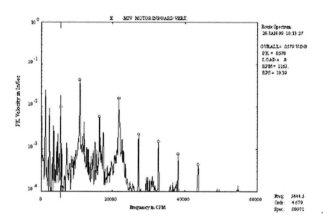

Figure 5.13   Enhancement of resonant energies viewed in *Y* log scaling.

According to the chart in Figure 5.14, we can roughly estimate the damping in the preceding plot to be approximately 0.2. The percentage of critical damping also can be calculated by saving a time domain of the ring down from an impact test. The ratio of any two adjacent time-waveform peaks can be used to calculate the log decrement curve and associated percent of critical damping. The excitation peak will always be high. Just consider the mound of energy that makes up the bottom 75% of the peak.

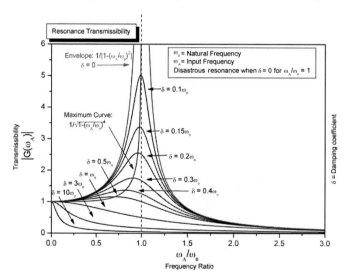

Figure 5.14   Different resonance responses due to percentage of critical damping.
(Source: Wikipedia.)

The user can calculate the damping ratio by taking any two adjacent time peak amplitudes and plugging them into the equation for x_0 and x_1 (Figures 5.15 and 5.16).

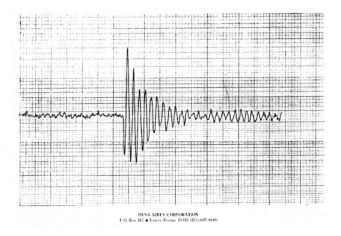

DYNA VIBES CORPORATION
P. O. Box 383 ● Exton, Penna. 19341 (215) 647-8660

Figure 5.15   Ring-down of a single natural frequency.

$$\zeta = \frac{1}{\sqrt{1 + (\frac{2\pi}{\ln(x_0/x_1)})^2}}$$

Figure 5.16   Formula to calculate the damping ratio from ring down.

The plot in Figure 5.17 was an impact analysis recorded on a two-pole induction motor. A clean natural frequency was identified at 3,300 rpm. This is sufficiently removed from the operating speed of 3,585 rpm to not cause any issues.

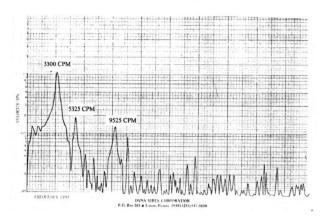

Figure 5.17   FFT response from impact analysis.

Induction motors create torque by rotor speed slipping against the field speed. The field rotates at 3,600 rpm, and the rotor rotates 3,585 rpm. Therefore, 2× running speed of the rotor would be 7,170 rpm. High spectral resolution is required to find these two frequency peaks that are just 30 cpm apart. The high-resolution spectrum in Figure 5.18 has shown that the predominant peak is not 2× running speed but rather 2× line frequency, or 7,200 cpm. This frequency is common in motors operating at 60 Hz because the entire stator core is excited by this frequency.

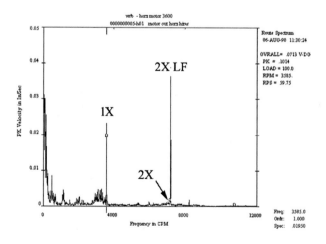

Figure 5.18   Need for high-frequency resolution.

Excessive vibration at 7,200 cpm in electric motors can be the result of numerous faults—for example, a rotor that is not centered in the air gap, a rotor that is vibrating in the air gap, or a loose stator core in the motor housing. Motor foot bolts can be loosened one at a time to see if the vibration drops. This is referred to as *elimination of soft foot*. For many years, the belief was that correction of soft foot was correcting for an uneven air gap. However, stators require a strong connection to the motor frame to minimize vibration at 7,200 cpm. If the motor foot bolts have been torqued such that the frame is twisted, the stator support can be compromised. Once a soft foot has been discovered, it should be additionally shimmed. Never loosen more than one foot at a time, always retorquing that bolt before moving to the next one.

C A S E   S T U D Y

A large water treatment plant had asked for a motor to be balanced on site. The motor was a large vertical motor and eddy-current drive resting on top of a pump head. The total height of the motor and pump head was 14 ft above the plant floor.

The customer wanted no vibration levels at the top of the motor in excess of 2 mils. That is a very low vibration level for such a high structure. The motor also was a variable-speed motor with an operating range of 400–900 rpm. Figure 5.19 is the first plot recorded on the top of the motor.

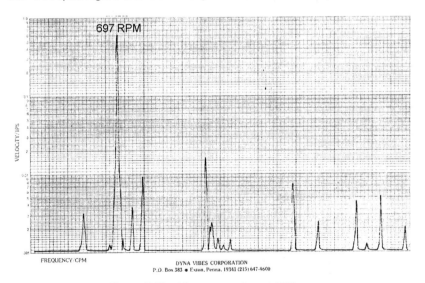

Figure 5.19　Motor operating at 697 rpm.

The customer described operation as variable speed, but the most constant speed range would be in the 700-rpm area. The vibration at the top of the motor at this speed was in excess of 0.5 inch/s, or just under 14 mils.

Note: Today's technicians can't appreciate how primitive the plotting capabilities were back when this case history was created. The signature was plotted on an *X,Y* pen plotter after the data were averaged and saved in the FFT. Frequency labels on the original plots were written at the time of capture by moving the cursor across the FFT screen and then using a ballpoint pen to write frequencies on the plot. I have taken the time to erase these original ink labels and replace them using tools in Adobe Photoshop.

A peak-averaged plot was recorded over the entire operating range to look for areas of resonance (Figure 5.20).

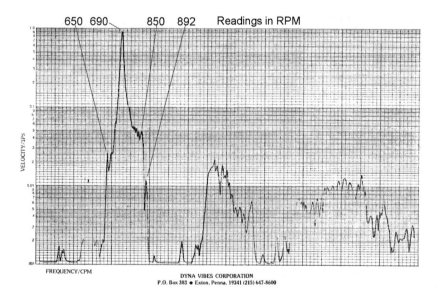

Figure 5.20   Peak-averaged plot over entire operating range.

This pump operated on an eddy-current drive. The constant motor speed was 892 rpm. The maximum output of the eddy-current drive placed the pump speed at approximately 860 rpm. Maximum motor vibration was just over 0.01 inch/s velocity.

The vibration level at the maximum pump speed was just over 0.05 inch/s velocity, or 2 mils. Notice how high the vibration grew as the speed passed through 690 rpm. Again, as the speed continued to decrease, the levels dropped off. Clearly, there was a resonance condition in the 700-rpm range, and 700 rpm was the speed at which the pump would operate for the majority of the time!

This case study is continued in Chapter 6.

REVIEW QUESTIONS

1. Electric motors generate heat in their rotor. What is the average time for a loaded motor to reach stable temperature?

2. A new induction motor is put into operation. The initial vibration is acceptable, but after 1 hour of loaded operation, the 1× vibration has increased to an unacceptable level. The motor is immediately uncoupled and operated solo, and the vibration at 1× is now unacceptable as well. What is the most possible explanation?

3. Trial weight runs have been completed, and the final weight calls for approximately 150 g. Using ¼-inch plate steel, what size piece should be cut if a ⅛-inch welding rod is to be used for attaching it? (Readers: Use the "Trial Weight Calculator" in Figure 5.4.)

4. A trial weight should generate a force equal to 10% of the rotor weight in order to establish the location and amount of imbalance. The rotor weighs 100 lb and has an operating speed of 3,600 rpm. What is the size of the trial weight in ounce-inches?

5. A newbie is out in the field getting ready to balance a fan. The vibration level is severe and predominantly 1× running speed in the radial direction. A trial weight is placed on the fan, and there is no change in the level or the phase reading. What is the problem?

6. The same newbie changes the trial and runs the fan for a new reading. The calculations call for the trial weight to be increased from 100 to 400 g. What should the newbie do?

7. True or False: Phase information is *always* required to balance a rotor.

8. True or False: Trial weights can be attached to the outer edge of a fan wheel just using the clamping bolt.

9. True or False: A rotor that has been installed in the field, experiences a rise in vibration levels and the ends are in phase. Eventually, the vibration levels call for the machine to be taken off line for being above operating specifications. The technician needs to balance this rotor by pushing the phase readings 180° from their original readings.

10. What setting should the technician use to capture the maximum vibration at every speed during a coast-down?

11. What setting can the technician change in the data collector or while viewing the data in the historical database to ascertain whether the reason for a large vibration peak is the proximity to a natural frequency?

12. True or False: The only thing that can cause a high vibration at 1× running speed is unbalance.

13. True or False: Balancing a rotor suspected of running near a resonant frequency calls for care when increasing the size of the trial weight indicated by a balance program.

14. The case study that involved balancing a gas turbine that could not pass through its first critical frequency called for use of the existing proximity probe outputs input into a digital FFT using peak-hold averaging. These results were then put through a no-phase balance procedure that resulted in a successfully operational gas turbine. What should the reader take away from this lesson?

CHAPTER 6

Machinery Fault Analysis, Part 3

Natural Frequency Analysis, Bearings, Alignment, and Process-Related Vibration

DOING AN IMPACT ANALYSIS TO CONFIRM NATURAL FREQUENCIES

We can now complete the basic trouble shooting charts by considering those items mentioned in the title. Correctly identifying a natural frequency problem requires the technician to learn some basics. It's just not about hitting a structure with a plastic mallet and watching the analyzer screen. Bearings, ball, roller, and sleeve all have their own unique appearance fast Fourier transform (FFT). Alignment, laser, and/or reverse dial is just the beginning. There are other issues that can affect even the best alignment, once the machine is loaded. Finally, the driven equipment has been engineered to operate under specific parameters. The end user must adhere to these, or unreliable performance will result.

Impacts have been discussed already. There are two major factors to be considered for doing them. First, the default window in your data collector is called a *Hanning window*. This is for analyzing periodic data or continuous data, such as monitoring vibration. A Hanning window ensures that there will be no erroneous data in the FFT. It does this by gradually zeroing out data at the beginning and end of every time window captured before the FFT (Figure 6.1).

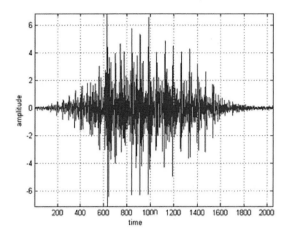

Figure 6.1 Periodic data after applying a Hanning window.
(Source: https://dsp.stackexchange.com/questions/18656/
reconstructing-time-domain-signal-with-hanning-window.)

During a capture of impact-response data, all the information is on the left side of the time window. Most analyzers have a window choice that will work for impact analysis. Choose a "Rectangular" window or "None." This will ensure that all the impact information is not zeroed out.

Notice the ring-down response (Figure 6.2). A Hanning window would remove more than 50% of the energy.

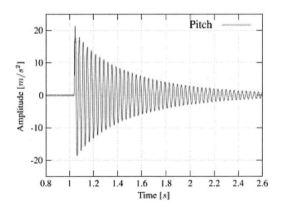

Figure 6.2 Typical ring-down response from an impact.
(Source: http://www.sciencedirect.com/science/article/pii/S0997754615300352.)

Also, the tool chosen to impact the machinery under test is important. If you are impacting a 2,000-hp motor, a large, soft-tipped dead-blow hammer should be used. If the hammer is small, bolt a steel mass to the opposite side of the tip. I was once asked to find the natural frequencies on a 4-inch-diameter turbine wheel. I used a 10-penny nail for the impact and a proximity probe for data collection! The frequency of interest called for a hard-tipped hammer. A large wooden timber suspended from a sling can be used to impact a large structure. This will work.

One of the scariest stories of excitation came from my first teacher, Ralph Buscarello. He was testing a four-story steel structure. He connected a steel cable from the structure to a bulldozer and had tension put on the cable. He then cut the cable with a cutting torch. This is *not* recommended. Mr. Buscarello has passed on, but he was single-handedly responsible for mentoring thousands of vibites with his infectious, enthusiastic approach to teaching the subject. Rest in peace, Ralph.

A previous employer of mine, Zonic Corporation, manufactured the ideal device for the preceding application. The product line is called *Xcite*, and the company is still in business. It sells what it calls an *inertial mass exciter* (Figure 6.3). It is a large steel box that contains a large mass on linear bearings. The mass is connected to hydraulic actuators that can move the inertial' mass back and forth at frequencies from 0 to 200 Hz. This device has been used by the U.S. Army Corp of Engineers to test earthen dams in Mississippi. That's a bunch of energy.

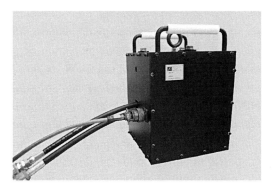

Figure 6.3 Xcite inertial mass exciter.
(Source: http://www.xcitesystems.com/1200-2inertialmass.html.)

CASE STUDY

Now back to impacting the tall vertical pump motor (Figure 6.4). Two highly responsive modes were discovered at the top of the motor. The lower frequency in the figure was perpendicular to the discharge piping, the less rigid of the orthogonal directions. Refer to Figure 6.4 to see how low the critical damping is.

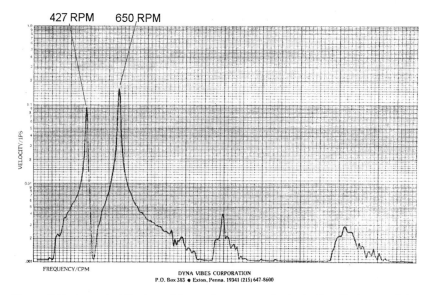

Figure 6.4   Impact at the top of the motor perpendicular to the discharge piping.

The frequencies are slightly different, but the response of the resonance in the 700-rpm range is clearly in line with the piping, whereas the lower-resonance peak (460–513 cpm) is perpendicular with the piping. This is usually the case because the piping adds stiffness to the in-line direction (Figure 6.5).

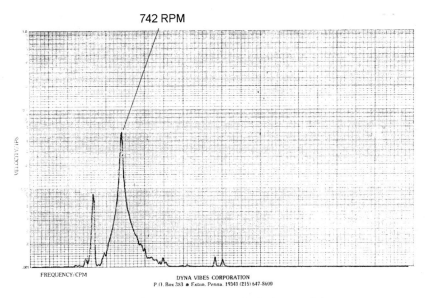

742 RPM

FREQUENCY/CPM

DYNA VIBES CORPORATION
P.O. Box 383 ● Exton, Penna. 19341 (215) 647-8600

Figure 6.5 Impact in line with discharge piping.

The customer was given the data to illustrate that the problem was not an imbalance. If the issue was just imbalance, the highest vibration readings would have been at maximum speed. In order to illustrate this concept, a 2- × 4-inch wooden beam was wedged between the top of the motor and the concrete wall. The beam was over 12 ft in length.

The plot in Figure 6.6 was captured in displacement for the benefit of the customer's specifications. The highest vibration during the coast-down was captured at the maximum speed, which is normal for just imbalance. The reading was 0.7 mil. The customer's first reaction was that the wood was holding the motor from vibrating and that eventually the motor vibration would hammer the wood out of place. I assured the customer that at these levels the beam would not move. I recommended trim balancing the top of the motor to eliminate the small residual imbalance of the coupled components, motor, eddy-current drive, coupling, pump shaft, and pump, as well as any off-center shaft alignment. I was asked to visit this site months later to commission additional equipment. The 2- × 4-inch beam was still in place on this pump.

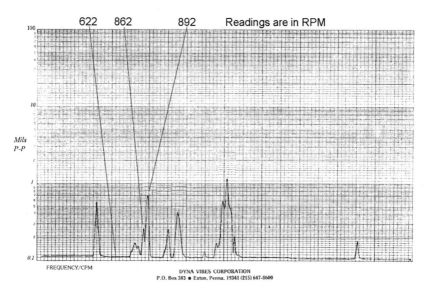

Figure 6.6   Peak-averaged plot over the entire speed range.

A 2,500-hp two-pole motor had been refurbished and returned to the customer, a large power utility. Abnormal vibration was detected (Figure 6.7), and a local consultant was hired to isolate the problem. He convinced the customer that the problem could be the rotor or the stator, so the customer purchased a replacement for both at a cost of tens of thousands of dollars.

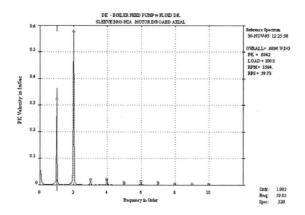

Figure 6.7   Vibration of large two-pole motor on a boiler feed pump.

Note: The consultant had used a portable proximity probe to analyze this motor. His conclusions were based on readings taken from reading the axial movement of the end of the motor shaft.

The customer called the motor repair shop that had refurbished the motor for a second opinion. Figure 6.7 shows the vibration on the motor non-drive end (NDE) in the axial direction. The 2× running speed is shown as almost 0.6 inch/s. The remaining readings on the motor locations indicated the same predominant peaks but at slightly lower levels.

The motor had sleeve bearings. There is no connection from the motor end bell to the shaft journal in the axial direction because the motor journal is resting on an oil film. The motor was coupled to the fluid drive with a gear-style coupling. Such couplings are very stiff and will not move axially once under load. So where did all the vibration go that was measured on the motor NDE end bell by the time it got to the fluid drive inboard bearing? The fluid drive bearing was a ball bearing. Any shaft vibration would be directly imparted into the bearing housing that was being measured (Figure 6.8).

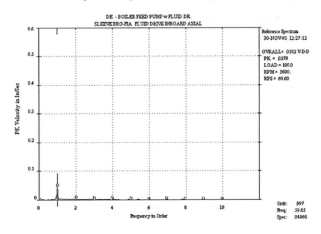

Figure 6.8 Vibration of the fluid drive coupling input bearing, axial direction.

The only conclusion possible was that the motor end bracket was vibrating axially, but the shaft was not!

The base of the motor 2× vibration peak appears to be broad. End bell resonance? The power to the motor was cut. As the motor slowed to approximately 300 rpm, the end bracket was bumped with a large, soft rubber mallet. Figure 6.9 is the captured spectrum from the impact.

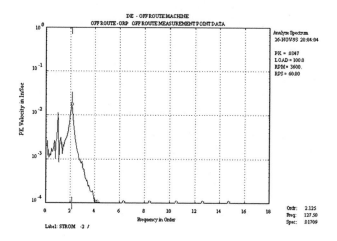

Figure 6.9   Axial impact on motor NDE bracket during slow roll.

The impact must occur before the motor shaft comes to rest. If the impact is performed while the motor shaft isn't turning, the stiffness of the motor rotor would be coupled through to the other motor bracket, yielding an incorrect result.

The test identified a lightly damped response in the motor end bracket at 7,650 cpm. The resonance is broad, and therefore, it is excited by the 2× line frequency of the stator. The customer wanted to know why the sister motor didn't have the same problem.

The motor bracket has a natural frequency that is 3 Hz higher than the number 1 boiler feed pump motor. This resulted in lower amplitudes (Figure 6.10), but it was still not acceptable.

The customer's engineer was distraught because the utility had purchased two expensive replacement components for no reason. The customer still wanted to know why he was seeing this vibration occur when it had not been there 3 months earlier. A graph on the wall, compiled by the customer for months, illustrated multiple parameters for this pump: load, running temperatures, vibration readings, and so on. The vibration readings were highest in the fall and spring and lowest in the winter and summer. The plant location was a marshy area near a major river. There was the answer! The land would dry out in the summer heat and freeze in the winter months. Both events added stiffness to the soil below the concrete foundation of the motors. Spring and fall rains would tend to soften the soil, thereby lowering the support stiffness below the foundation.

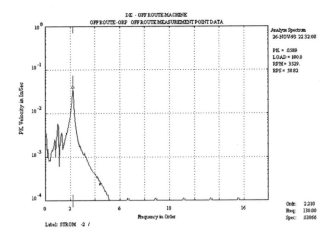

Figure 6.10   Impact on second boiler feed pump motor end bracket.

The report conclusion to the customer was that the natural frequency in the end bracket existed in the original design and, because there was no stress to the bearing or shaft by the bracket resonating, to let it run. When asked how long the motor had operated before the refurbishment, the customer replied, "Twenty years." That motor is still running, and this job was completed in 1993.

I had just joined Reliance Electric and was not familiar with CSI MasterTrend software or the 2115 data collector. William B. Fagerstrom, of the Engineering Department of E.I. Dupont de Nemours and Company, was instrumental in the development of the software and the 2115 data collector specifications. He graciously agreed to accompany me, on Thanksgiving Day, to the job site. While looking at the vibration data, Bill quietly set up the 2115 for an impact data capture. Then he asked plant personnel to cut power to the motor. The motor slowed, the motor bracket was impacted, and the rest is history! Dr. Fagerstrom still teaches a course titled, "Manufacturing and Processes and Systems," at the University of Delaware.

CASE STUDY

A military facility remanufactured helicopter jet engine gearboxes. The technicians were getting different test results on the same gearbox when it was tested in two different cells. The facility also was interested in purchasing a FFT. I inspected the "good" cell components. The gearbox was the key component for pass/fail. The gearbox original equipment manufacturer (OEM) had manufactured the test bracket for mounting the accelerometer. The bracket was mounted using two of the gearbox access cover bolts. It was also machined as a tuned bridge to be resonant at the gearbox primary gear mesh frequency. The accelerometer was a high-frequency model, and the output was fed to an overall meter. The output also was fed through a filter to a pass/fail light. The data from a running gearbox were fed into the FFT, and the spectrum was plotted. Then the accelerometer, mounting bracket, and cable were brought from the "bad" test cell and put on the same engine in the "good" test cell. The reading on the overall meter was the same. However, the energy on the FFT plot was now centered almost 50,000 cpm lower! The customer was now totally confused. The inspection of the mounting bracket revealed the answer. The operator of the "bad" cell did not want to be bothered taking the two mounting bolts completely out of the gearbox cover to mount the test bracket, so he had cut slots into the holes of the test bracket! This had changed the natural frequency of the bracket. The customer bought a FFT the next week, making the FFT salesperson, who had set up the visit, very happy!

BENT SHAFT

Bent shafts can be a cause of high vibration levels at 1× running speed or 2× running speed depending on where the bend is. A bend near the midpoint of a rotor shaft will appear as an imbalance. A bend near a bearing journal will cause a high 2× component because the bearing will appear to be misaligned or cocked on the shaft. Multiple vibration and phase readings are required to pinpoint this problem.

Warning! Attempts to reduce vibration levels resulting from a bent shaft may take a considerable mass. If the item in question is a fan, for example, a very large trial weight could end up pulling the fan apart. I have found that run-outs on the order of 0.005 inch can be dealt with. Run-outs in excess of this require shaft replacement.

SLEEVE BEARINGS

Antifriction bearings have been discussed in this book, but sleeve-bearing problems also exhibit unique vibration characteristics.

C A S E   S T U D Y

The machine in this example was a chlorine gas compressor. The compressor was driven by a two-pole motor through a gear increaser. The compressor speed was 8,400 rpm. There were multiple peaks of subharmonic frequencies. The multiple peaks were based on a fundamental frequency at 3,300 rpm. This is less than 50% of running speed and indicative of a worn sleeve-bearing bore (Figure 6.11). Wear within a sleeve-bearing bore area causes problems with the formation of the supporting oil wedge (Figure 6.12).

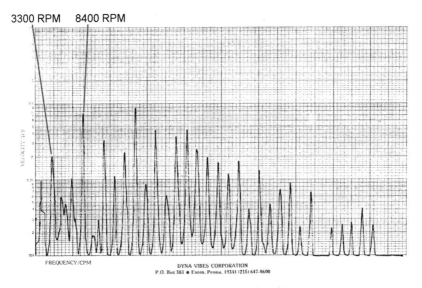

Figure 6.11   Worn sleeve bearing.

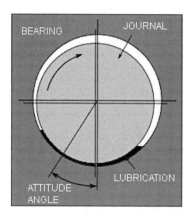

Figure 6.12 Location of proper oil wedge formation.
(Source: Wikipedia.)

The customer was informed of excessive clearance in the sleeve bearings. The sleeve bearings were sent out for recasting. The new bearings were scraped in by a seasoned machinist for proper clearance. The bearing clearance was specified to be between 0.004 and 0.006 inch. The machine was reassembled, and vibration readings were recorded (Figure 6.13).

4200 RPM   8400 RPM

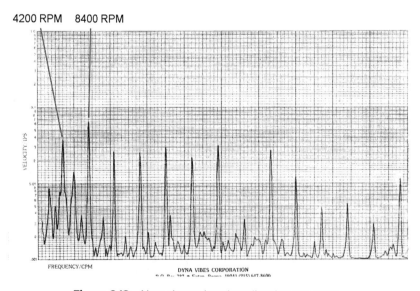

Figure 6.13 New sleeve-bearing vibration spectrum.

The new sleeve bearings had initial vibration readings that were not acceptable. The predominant energy consisted of peaks at exactly half the running speed plus harmonics. The technician asked about the bearing clearance and found that the bearing clearance was left at the minimum of 0.004 inch. A recommendation was made to open the bearings another 0.001 inch. Management fought this, explaining that specifications were specifications. After one whole hour of arguing, the machinist was permitted to open the bearings another 0.001 inch.

The final vibration readings indicated that the oil wedge was properly formed, the rub was gone, and vibration was well within satisfactory limits (Figure 6.14). Vibration at the compressor 1× running speed of 8,400 rpm was just over 0.03 inch/s velocity.

8400 RPM

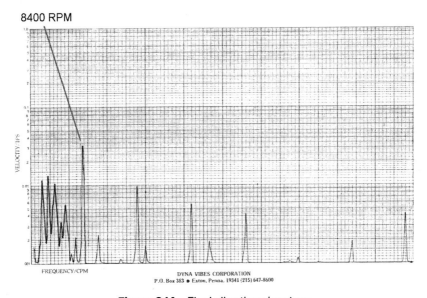

Figure 6.14   Final vibration signature.

A major utility asked me to visit its facility where a recently reassembled steam turbine generator was operating at vibration levels in excess of 0.3 inch/s velocity on the high-pressure stage. The vibration readings indicated multiple half-times operating speed harmonics in the first-stage turbine bearings. The plant manager discussed the work just completed, and records indicated that the clearances in the bearings had been left at the minimum specification. The turbine bearings were on a closed-loop oil lubrication system with the bearing temperatures and oil lube collection tank all closely monitored. The suggestion was to raise the oil lubrication temperatures while ensuring that bearing temperatures remained within specifications. The hope was that this small change might allow the proper oil wedge to form and eliminate the internal rub. The operators agreed that they could raise the temperatures approximately 5°F and still be within specifications. The cooling water to the oil lube holding tank was throttled back as all temperatures remained monitored. When the oil lube temperature and bearing temperatures reached a point 4°F above the original temperatures, the vibration level dropped within milliseconds from 0.3 to 0.03 inch/s. Set points were changed on the temperature monitors, and the utility proceeded as normal. A procedure also was implemented to increase cooling water at the moment of a shutdown to maintain temperature specifications.

A pure sine wave contains one frequency. The addition of a second frequency results in a combination of the two. The spectrum in Figure 6.15 identifies 1× and 2× running-speed vibration on a large sleeve-bearing motor. There is no impacting because of sufficient clearance in the bearing, so there are clean 1× and 2× components that are evident in the time waveform.

The plot in Figure 6.16 was recorded on the pump in the axial direction. The predominant frequency is 5× running speed and most probably due to vane pass frequency.

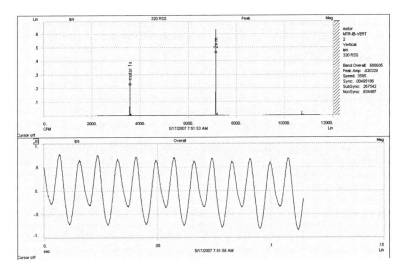

Figure 6.15   Example of 1× and 2× vibration spectrum and time waveform.

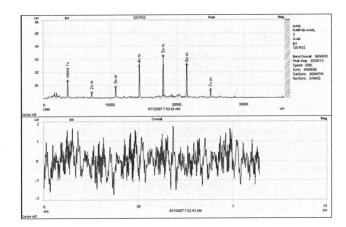

Figure 6.16   Higher-frequency content in the time waveform and spectrum.

Proximity probes are an excellent tool for diagnosing sleeve-bearing issues. The instrumentation should include two orthogonally placed probes and a key phasor for direction of rotation and speed.

Another problem that can be encountered with sleeve bearings is referred to as *oil whip*. This occurs on machinery that operates above the first bending mode of the shaft. As the machine passes through that critical frequency, there is a tendency for the bearing and the oil wedge to become unstable. Under certain conditions, oil whip can result. The natural frequency will be excited, and then, even though the speed continues to increase, the shaft resonance remains, and very high vibration levels result.

BALL BEARINGS

Ball bearings were discussed earlier.

CASE STUDY

This case history reinforces the need for precision when installing rotating components. A paper mill had replaced bearings and shaft on a hot-well pump. The new bearings had been in operation for only a week, but they already showed signs of failure. The unit was removed and brought into the shop for inspection. The bearing bores and shaft journals were measured. The shaft journals were found to be 0.002 inch oversized. The bearings were a shrink fit. When they cooled, this oversized condition had preloaded the ball bearings. The specified clearance in the ball bearing had been reduced by that 0.002 inch, and the bearings overheated rapidly. The shaft was turned down to specification, and new bearings were installed.

This analysis was performed using new equipment the customer had bought from Nicolet Scientific. After the successful audition, the plant manager wished to speak with me about just how this new equipment worked. Because I was accustomed to giving vibration training seminars, I went into the explanation with my typical enthusiasm. Pointing to the high bearing energy, I explained that this was a clear indication of a "bearing going South." I stopped dead in my tracks. This paper mill was in Mobile, Alabama! The plant manager just smiled and replied, "We like to say the bearing is going North." Then he grinned ear to ear and politely asked me to continue with my lesson. Shrink fit procedure calls for measuring the bearing bore and the shaft journal. The difference in measurements should be 0.0005 inch, per inch of shaft journal diameter.

The before and after plots are in Figure 6.17.

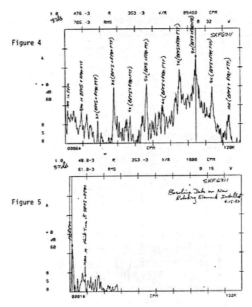

Figure 6.17 Before and after ball bearing replacement and shaft journal machining. (Vibration signatures courtesy of James S. Stinson, Scott Paper, Mobile, Alabama.)

C A S E   S T U D Y

Warranty Administration sent me to investigate a vibration problem with a newly delivered dynamometer. The dyno had been assembled in Michigan and shipped to Riverside, California. The dyno was built around a new pancake 400-hp motor-generator that was designed to fit between the driven-wheel span of an automobile. The output shafts each had 1,000-lb rotors that supported the automobile's wheels.

A run-up of the dynamometer sounded bad. There was something wrong. A Zonic WCA analyzer was used for data acquisition. This test equipment had up to 32 input channels and the ability to record all data with a real-time rate up to 5 kHz. This real-time rate describes the frequency range the technician can capture with zero gaps in the data. In other words, if there is a transient spike, it will be captured. The analyzer had an excellent capability that was the result of input from Detroit automotive engineers. The data could be played back with dedicated memory set to record given orders of running speed. The data were all saved as order-tracked data, meaning that the internal clock of the FFT is bypassed, and the data-collection timing is slaved to an encoder reading shaft speed. Dedicated orders to be saved included 1× running speed, ball pass frequency outer (BPFO), ball pass frequency inner (BPFI), and ball spin frequency. The results are plotted in Figure 6.18.

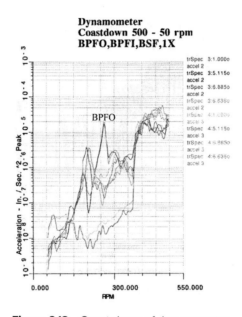

Figure 6.18   Coast-down of dynamometer.

The two large peaks in the center of the plot (identified by "BPFO") represent the ball pass frequency outer race from both motor-generator bearings. The amount of vibration on such a slowly rotating rotor was cause for concern. There was a serious defect in the outer race. Both bearings had cracked outer races! I made a call to the manufacturer of the dyno. The question was asked, "Did you ship the entire dyno with the output shaft rolls in place?" The answer was affirmative. "Did you block support the rolls?" The answer was negative. The manufacturer's representative claimed that an analysis of the bearings indicated that they could support the roll static weight without damage. I replied back, "Until the truck hits a pothole. The dynamic loading on the bearing will then go up by an order of magnitude." I could see that person on the other end of the line was waking up. He exclaimed, "Oh man! We were getting ready to put eight of these units on a ship to Korea." This one failure and the lesson learned had saved the company from eight damaged units. The bearings would never have survived an ocean crossing.

Note: The bearings were unstable until the speed dropped below 375 rpm. The 1× vibration on both bearings dropped significantly, but BPFO amplitudes were 100–1,000 times higher.

Another failure mode for ball bearings is referred to as *brinelling*. There is true and false brinelling. *False* brinelling is the result of the ball or roller polishing out the finishing grinding marks on the inner and outer races. This also includes corrosion damage that results from the ball/roller motion removing the protective lubrication coating. *True* brinelling damage occurs when there is a tactile displacement of the bearing material. A bearing with false brinelling but no corrosion, by my definition, can yield a normal life because the finishing grinding marks are polished away during normal bearing use.

The final bearing condition in Figure 6.19 is the result of a bearing being dropped prior to installation or the assembled equipment being subjected to a high-*g*-force impact, possibly during shipping and handling. The balls damaged the inner race before the bearing ever made one revolution. Rapid failure of the bearing occurred as more and more metal was ripped from the race. The spacing of the circular damaged areas is the ball spacing as supported by the cage. This is the result of true brinelling damage on a new bearing.

Figure 6.19   Example of spalling as a result of true brinelling.

CASE STUDY

A manufacturer transported 30 motors for pump applications at a power plant. The motors had been shipped with 100-lb sheaves mounted on the motor shaft without blocking. The warranty administration was looking at the potential of having to replace 30 sets of motor bearings. The technician sent to the field to assess the damage took high-resolution FFT and time-domain data on all 60 bearings. The testing resulted in only five of the motors requiring new bearings.

BALL BEARING INSTABILITY

CASE STUDY

A large OEM had historically purchased all alternating-current (AC) motors with sleeve bearings. The decision was made to lower costs and change to ball bearings. One of the first ball-bearing motors was experiencing high vibration levels while in test. The motor was mounted on the equipment but not coupled with the driven equipment. Vibration levels were in excess of 0.3 inch/s velocity.

The sidebands were measured at approximately 240 cpm spacing. Nothing in the operation of this motor was thought to generate this frequency (Figure 6.20).

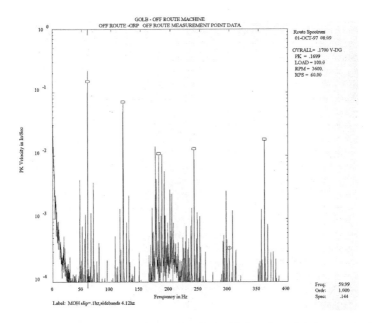

Figure 6.20   Running vibration with a high-sideband-modulation content.

A coast-down of the motor was initiated to capture the vibration data. The coast-down uncovered the fact that the 1× vibration levels did not begin to drop until the motor had slowed to 2,880 rpm! The ball bearings were not tracking until that speed was reached. The excess spacing in the bearings caused speed modulation. The cage would speed up as a ball passed through the load zone, but it would then slow down as the balls were on either side of the load zone (Figure 6.21). Bill Subler, chief mechanical engineer at Reliance Electric Large Motor Facility in Kings Mountain, North Carolina, diagnosed the problem. The bearings were preloaded with the addition of Smalley washers (wavy spring washers). This additional stiffness forced the balls to track properly on the inner and outer races.

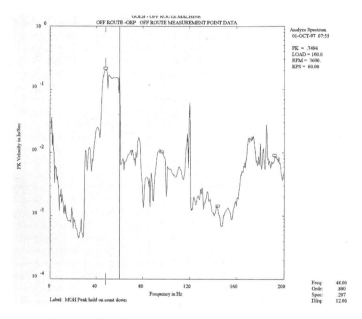

Figure 6.21   Coast-down peak hold—averaged plot.

LOOSENESS

Looseness is another common machinery fault that is easily diagnosed. Early in this book, a typical vibration signature was defined as having four or five harmonics of the fundamental peak. Looseness is most readily identified by the existence of multiple harmonics, sometimes as many as 20 or more.

CASE STUDY

The data in Figure 6.22 were recorded on a steam turbine that had a normal operating speed of 5,000 rpm. The turbine could not get above 3,000 rpm without severe vibration issues. The 1× vibration was in excess of 0.3 inch/s. The turbine had been recently overhauled, so the decision was made to perform an internal inspection. The turbine wheel assembly was not shouldered. There was a gap between the turbine wheel shoulder and the mating face on the shaft. The turbine wheel assembly also was sitting on a tapered shaft. The assembly was removed, and the wheel assembly was correctly seated.

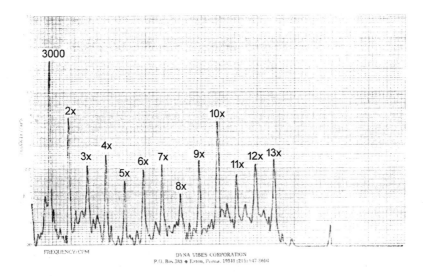

Figure 6.22   Looseness within a steam turbine rotor.

The final vibration reading of 0.055 inch/s *and* the removal of the multiple harmonics of operating speed indicate that the assembly is correctly seated (Figure 6.23).

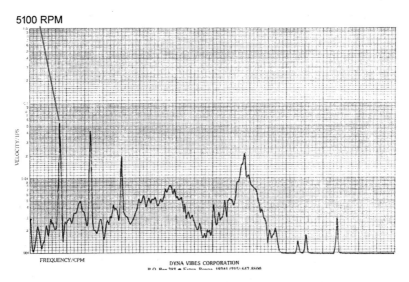

Figure 6.23   Final vibration on steam turbine at 5,000 rpm.

MISALIGNMENT

Following balance, the most important component to ensure smooth operation of rotating machinery is alignment. Today's technicians have the benefit of the laser to assist in this task. Machinists were historically trained to use dial indicators for this task.

Reverse dial alignment is the best method for determining where two shaft centerlines are in relation to each other, as opposed to "rim and face." (Figure 6.24).

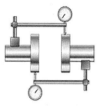

Figure 6.24   Mounting of dial indicators for reverse dial alignment.
(Source: https://engineeringcontent.blogspot.com/2014/10/
misalignment-costs-time-and-money.html.)

The results of the dial readings are then plotted on graph paper. There will be two graphs, one for the vertical alignment results and one for the horizontal results. The graph in Figure 6.25 indicates that the motor vertical misalignment is due to the rear motor feet being 0.030 inch too high. Notice that there is no call for additions to or subtractions from the front feet. The motor shaft centerline will pivot through this point as the rear feet are lowered.

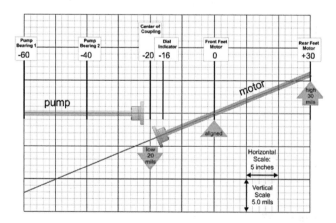

Figure 6.25   Graphical representation of vertical misalignment of motor and pump.

Laser alignment tools have reduced machinery alignment times drastically. The laser head and reflective prism are easily and quickly mounted to the shafts. Global positioning in the laser head automatically identifies where the laser is as the shafts are rotating. The shafts are turned together, meaning that the coupling doesn't need to be disassembled for the procedure (Figure 6.26).

Figure 6.26   Modern laser alignment equipment.
(Optalign Laser Alignment Tool, Ludeca.)

Alignment of twenty 600-hp, 3,600-rpm compressors was requested by a customer. All units were 10 ft above ground level, sitting on fabricated steel framing. All motor foot bolts were through bolts, meaning that another person was needed below to hold a wrench on the nut. The two back motor feet bolts were at the back edge of the platform behind the motor, requiring the technician up top to use his foot on the wrench to break the bolts free. Because of labor rules at the site, there were two 30-minute breaks and a 1-hour lunch break.

The job was completed in three days. The previous alignment had been performed using dial indicators and the *rim and face method*. The rim and face method is notorious for leaving incorrect shimming on the rear feet. This is due to insufficient accuracy for the face reading, and it is usually performed with the coupling open, so run-out on the face also will cause inaccuracies.

A stack of more than 7 inches of shims that had been removed from the rear feet was handed over to the customer at the completion of the 20 alignments.

Now for the pitfalls: I discussed soft foot earlier. Most alignment tool manufacturers instruct users to tighten all the feet first and then loosen one foot at a time and record the readings. The result will indicate which foot or feet require additional shimming. Unfortunately, this method assumes a rigid base. Many motor foundations in the real world are less stiff than the motor frame. This makes for erroneous readings when checking for soft foot. The base is coming up to meet the motor foot rather than the foot being tightened down to the base.

A method for eliminating soft foot that has worked well for me is the following: Remove any old shims, and make sure that the bottoms of the motor feet and the mounting base areas are clean.

Depending on how low the motor is, to start, place the same thickness shim under all four feet. If the motor needs to be raised more than ¼ inch, then ¼-inch plates should be cut to the foot size and used as a starting point. Lower the motor down onto the shims. Check each shim by hand to see if it will move or spin. Any shim that can be moved by hand indicates a foot that requires additional shimming. Remember, there should be no more than four or five shims total under any one foot. More than this number will result in what is known as a *squishy foot*. This results in difficulty getting the final vertical readings quickly.

Once the shims have been added and no shims can be moved by hand with just the motor weight resting on them, proceed with the vertical alignment procedure. Always use new, clean stainless steel shims. A good rule of thumb is to move the motor vertically a few thousandths of an inch higher than a zero reading. Then, as the motor feet are tightened, the clearance between the shims will be removed. Tighten the motor feet as you would the head on an internal combustion engine (Figure 6.27).

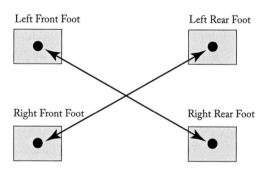

Figure 6.27   Torqueing foot bolts.

Move diagonally and watch the reading on the laser output. The reading will be above or below zero. As one motor foot bolt is tightened, the reading will approach zero. Take the reading a couple of thousandths of an inch to the other side of zero and then move to the foot located diagonally from the present foot, and start to tighten that foot bolt. Again, the reading will approach zero and swing through to the other sign. Keep switching feet, and continue to torque down the foot bolts. Eventually, the reading will no longer move. Take a set of readings with the laser to see what the vertical alignment is. It should be very close to specifications.

The side-to-side alignment is much easier, with live readings as the motor is moved. You will have to reduce the torque on the motor foot bolts so that side-to-side movements can be made. With practice, both alignment moves can be accomplished simultaneously.

There is one key piece of information that will never come from the laser alignment tool: the axial distance between the two couplings. This is critical because a distance that is greater or less than the coupling specification can cause an axial preload on the fixed ball bearing. This information is even more critical for a sleeve-bearing motor. Sleeve-bearing motors normally do not have thrust bearings. This means that the driven equipment must carry the total thrust load. This makes sense because there is no thrusting force in an electric motor. The rotor has a shoulder on each side of the sleeve bearing. It is critical that the axial position of the motor rotor and frame do not cause this shoulder to ride on the side of the sleeve bearing. Should this occur, the bearing will melt in minutes on startup.

Every sleeve-bearing motor shaft should be marked with three lines: magnetic center (the position of the motor shaft when running unloaded, uncoupled) and the total mechanical end float movement of the shaft in the bearings. Ideally, magnetic center should be centered between the two mechanical end float stops. Some motors are engineered with removable shims between the bearing inserts and the bearing housing. This allows plant personnel to adjust the liner axial position to achieve placing the magnetic center at the midpoint of the end float positions. If this cannot be accomplished, the technician is better off placing the motor shaft such that the end float is equal in each direction. Running near any of the bearing shoulders raises the risk of wiping out the bearings. Running the motor with the rotor off magnetic center is of no consequence. The axial force in an electric motor due to running off the magnetic center is minimal and will cause zero degradation of machinery components. Some older motors have such large air gaps, that the uncoupled rotor will drift to one end if the motor is not level.

Falk- or gear-style couplings have float. This float must be limited by using a limited end float button inside the coupling hubs. If these are not used, the motor shaft can move axially, and the motor shaft shoulder will wipe out the bearing. The button is a steel disk that fits inside a rabbet fit in the coupling shrouds. If a spacer is part of the coupling setup, there will be two buttons, one for each side.

Therefore, once the coupling has been assembled correctly, the first step for the technician is to move the entire motor frame axially, forward, or backward until the magnetic centerline is properly set to the reference mark on the bearing housing. This should be clearly marked and is sometimes indicated as a measurement from the seal face of the inboard bearing housing (Figure 6.28).

Figure 6.28   Falk coupling (gear).
(Courtesy of Falk Coupling, Inc.)

If any part of the alignment cannot be achieved because the foot holes are bolt bound, the technician has two options:

1. Slot bore the motor foot holes.
2. Neck down the shank of the motor foot bolts.

Necking down the shank of the bolt will require some engineering to ensure that the torque levels required to hold down the motor will not shear the bolt.

There will come a time when the alignment readings indicate that the rear motor feet must be lowered and there are no shims under the feet! The first time I encountered this, the laser was reversed to see what the results were for the driven equipment. Same thing! The back feet needed to be significantly lowered, and there were no shims! The manual was no help, so a call to tech support was in order. The answer made sense. Add shims to the front feet of *both* the driven and drive equipment! That brought the two pieces up into alignment with each other. The rest of the alignment was straightforward.

CASE STUDY

This is an example of a machine operating with approximately 0.250 inch of shimming required on both inboard feet of both machines to bring them up into alignment. The machine was used to crush large tree trunks (up to 20 inches in diameter). Notice that the "Corrections" section is calling for removal of 0.434 inch from the rear feet (Figure 6.29).

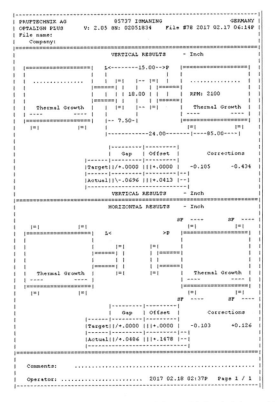

Figure 6.29   Alignment sheet on machine with both inboard feet low.

Warning! The previously referenced alignment procedure may not work on equipment with rigid piping connected, such as pumps, compressors, and so on. Such equipment doesn't usually have this problem unless the initial installation was not correct. There are smaller pumps where the piping bolts on the inlet and discharge flanges can be loosened until the alignment is correct. This has been done on new installations with success.

THERMAL GROWTH

Pity the poor technician. The amount of thermal growth a machine will encounter is discussed but very seldom are the actual numbers known. This is due to several factors. The manufacturer may know the growth numbers when the machine was tested in the factory, but on site the manufacturer has no control over the foundation or the piping. Frequently, the piping can cause growth restrictions or be the major cause of forces in the machine if not hung and isolated properly.

A customer specified that large electric motors should be set 0.005-inch low for thermal alignment. It is my opinion that there are no real thermal growth issues with electric motors. There was no difference in the vibration levels of the motors that were set line to line (0.000 inch) or 0.005-inch low. The main reason was that they all ran with spool-piece couplings that were anywhere from 12 to 24 inches long. The addition of any spool piece allows for additional parallel misalignment at the rate of 0.0003 inch per inch of coupling spool. A 12-inch coupling allows for the parallel alignment to be off by 0.004 inch! See Figure 6.30.

Figure 6.30   Complete coupling, hubs and spool piece.
(Courtesy of VibrAlign, Inc.)

There are some machines that have such high rates of compression or expansion that one end of the compressor is too hot to touch and the other end grows ice. This is a machine that should have its thermal growth numbers considered. Again, the manufacturer is a good place to start for recommendations. If information is lacking, the technician can begin by understanding that the end of the machine that is cold will not grow at all. The hot end will grow, so that end should be set low. Also check the machine mounting. Some machines, in order to minimize or even eliminate thermal growth issues, have bases designed to support the machine from the bearing centerline. Then the growth will be even above and below the bearing centerline.

Note that the inlet and discharge pipes in Figure 6.31 are both vertical and isolated from fixed piping by flexible bellows. The thermal growth of the housing still must be accounted for.

Figure 6.31   Hoffman blower.
(Courtesy of Hoffman Blower, Inc.)

Some OEM manufacturers have spreadsheets to assist maintenance personnel in setting a known thermal growth pattern. However, their calculations are based on the foundation and piping being correct. The end user reads the temperatures on various parts of the equipment and places those readings into the spreadsheet to yield recommended alignment offsets. Alignment tools have the means to program these offsets into the tool so that the user can then just align to zeroes.

Ray Dodd, an engineer working for Chevron Oil Company in the 1970s, came up with an ingenious method for measuring and trending thermal growth. The system consists of two triangular framed bar sets rigidly mounted to the drive and driven equipment. The bars were manufactured from Invar, a composite metal known for its unique low thermal coefficient of expansion (Figure 6.32). In other words, the growth changes measured would not be due to the bars. One bar was equipped with two clamps. Each clamp held two pairs of proximity probes (remember them?). The other bar held two steel target blocks for the probes to read. The probes were set and zeroed when the machine was cold. The output of the probes was read on a digital multimeter or recorder. The machine was started and brought to full load. Once normal operating temperatures were reached, the output of the probes was plotted on graph paper to find the final thermal growth numbers. The technician could then put these offsets into the cold alignment, and the machine would grow to an alignment well within specifications.

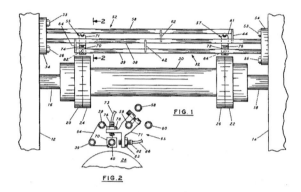

Figure 6.32   Dodd alignment bars.
(U.S. Patent No. 3783522, Ray Dodd.)

CASE STUDY

I have had experience with these bars, and they work well. I once set up a pair on a steam-driven propane compressor. Probes were zeroed and left for the overnight run. When the technicians returned the next morning, the probes were crushed and the bars bent. The mechanics, working on the overhead piping, had used them as a ladder.

New equipment was installed and the test rerun. The readings called for the front feet to be positioned 0.100-inch lower when at ambient temperature. The bolts on the front feet were loosened, and the shims fell out! This was because the front feet were off the foundation! The vertical discharge piping was holding the entire machine up. A *Dutchman* was machined to place in the first vertical piping flange. This brought the machine down to the foundation, and a correct cold alignment was achieved. Dutchman refers to a spacer that is machined with the proper face angles to be used as a spacer without cocking the piping.

CASE STUDY

A power utility customer complained about a gas turbine that never ran consistently at acceptable vibration levels. Large gas turbines have very wide temperature variations. This turbine shaft grew approximately 0.150 inch in length from cold to hot. The manufacturer had designed the turbine supports to allow for this growth while ensuring that the turbine-generator shafts remained in line.

The customer's main complaint centered on the fact that one start might yield good vibration levels and the next start would have higher vibration levels. The unit would be stopped with an immediate restart, and sometimes the levels would begin to drop back down. Other times, the readings were still rough or sometimes even higher. The frustration was palpable, with zero help from the manufacturer.

Thermal alignment changes were suspected, but how was one to measure such movement? The decision was made to go with four laser measurement heads to monitor the horizontal movement of the two main turbine support trunnions and the axial movement of the two exhaust support trunnions.

The model selected has a total range of 0.5 inch, much greater than a proximity probe. The sensitivity of the measurement is 0.00015 inch, and the frequency range for dynamic measurements is 1,250 Hz (75,000 rpm). Because the sensing mechanism is a laser beam, the target can be any material (Figure 6.33).

Figure 6.33   Acuity displacement laser head.
(Laser displacement sensor; available at
https://www.acuitylaser.com/products/item/ar200-12.)

The customer had his answer within days. The movement of the trunnions was proven to be drastically unequal. The north turbine trunnion (measuring horizontal movement) moved 0.023 inch, while the south turbine trunnion moved 0.003 inch. The north exhaust trunnion (measuring axial movement) had no movement (0 inch), while the south exhaust trunnion moved 0.015 inch. The shafts were being forced into a bow by the irregular and unpredictable movement of the housings (Figure 6.34).

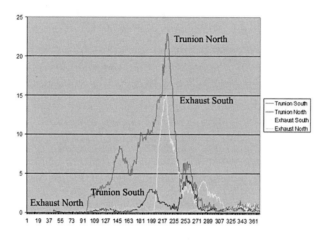

Figure 6.34   Acuity laser results from large gas turbine run.

The customer proceeded to rebuild all four trunnions. The results had narrowed it down to the north exhaust trunnion, which had zero movement. This one trunnion was frozen.

C A S E   S T U D Y

A very large database center was experiencing cracks in the bell housing that connected the generator to the diesel engine. Vibration was suspected, but vibration levels were not excessive. Figure 3.35 illustrates the setup of the lasers to measure the rails below the generator.

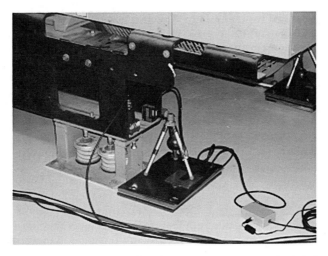

Figure 6.35   Acuity laser head setup to measure vertical static
and dynamic motion of DG rails.

The vibration of the left and right rear rails are identified in Figure 6.36.
They identified a 0.020-inch opposite-direction vertical transition of the rails.
The isolators below the rails were correct for the application, but their total
travel was supposed to be limited by adjusting snubbers. The snubbers had
not been brought up to limit travel. As a result, the movement allowed by
the torsional loading and unloading of the generator placed an excessive
torsional twist on the generator bell housings, and they cracked.

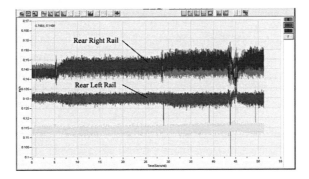

Figure 6.36   Laser readings from the four rails of the Diesel-Generator.

BELT ALIGNMENT AND BELT TENSION

C A S E   S T U D Y

I was asked to investigate the failure of multiple motors operating on presses at an automotive manufacturing plant. The motors had all failed in the same manner. The inboard roller bearings had been destroyed. The rollers had been worn from round to semicircles!

The motors were the main drive to the flywheel on the press. The flywheels were 6 ft in diameter. The belts were 8V × 5. This is a solid belt with five grooves. These belts are substantial. A test by hand of the belt tension proved that it was extreme. The customer stated that a new device, an acoustic belt-tensioning gauge, had been brought in to test them. The conclusion was that the belts were too loose! The customer was asked to have a mechanical belt-tensioning gauge on site the next morning. Meanwhile, vibration data were recorded on one of the newer motors. The ball pass frequency of the outer race was visible in the data on a brand-new motor. The balls were being squeezed through the load zone!

An inspection of the motor installation found that it was supported by a solid-steel 2-inch-thick plate. The site mechanic exclaimed how difficult it was to set the belt tension. He claimed that the technicians had started with wrenches, and when they could not get the torque required, they used 10-ft pipes on the wrenches. When they still could not reach the OEM-recommended belt tension, they used a 2.5-inch-bore hydraulic ram under the 2-inch mounting plate (Figure 6.37). I asked where the belt tension specification had come from, and what was it? The mechanic grabbed a paper from his folder, and it showed that the belt tension was to be set by a measurement between the flywheel hub center and the motor shaft center, not by measuring tension!

Note: Notice the size of the jacking bolts. They were 1.25 inches in diameter.

The mechanical tensioning gauge was on site the next morning. The formula called for a side movement on the belt of approximately 1.5 inches for the correct tension. The operator climbed up inside the flywheel well and pressed the gauge against the belt. When the gauge reached the proper tension reading, the belt had moved 0.125 inch (Figure 6.38).

Figure 6.37   Solid-steel plate beneath motor—2-inch adjustment bolts.

Figure 6.38   Flywheel with belts—location of belt tension test.

I asked the customer to reset all the motors to the correct belt tension and flew home. Both the plant manager and I waited for an answer from the OEM on the belt-tensioning dimension. The answer came back in several weeks. The dimension was off by 3.5 inches!

Always use a belt-tensioning gauge. By the way, acoustic belt-tensioning tools have been perfected and work well.

PROCESS-RELATED VIBRATION

CASE STUDY

A chemical manufacturing facility lost a 600-hp motor. The motor was sent to the motor repair shop, and the damage was so extensive that a new motor was purchased. The new motor was installed and put into operation. The customer checked all operating parameters and signed off.

I was called by the facility at 3 a.m. the next morning. The unit was exhibiting high vibration. A site visit proved the issue to not be repeatable. All was well with the unit 1 hour later. I left.

A call came into the motor repair shop the next day. The unit was acting up again. Again, by the time I arrived, the problem was gone. The electrical maintenance manager was asked to install a current probe and a chart recorder and call when the "symptom" was captured.

The next Monday, the manager called and said that he had something he wanted to fax to my office. The night before (sure, always on a Sunday night, yes?), the chart recorder captured an issue with the motor current.

The current fluctuations that had been captured were severe. Approximately 50% of the load was being picked up and then dropped in fractions of a second. The customer described the chiller operation. The full load of the chiller was not required for the process, so a bypass valve had been placed in the discharge line to take 50% back to the supply side. The check valve in the bypass line was malfunctioning. It was opening and slamming shut numerous times in seconds. This was the cause of the original motor failure, and it would have resulted in failure of the new motor in no time (Figure 6.39).

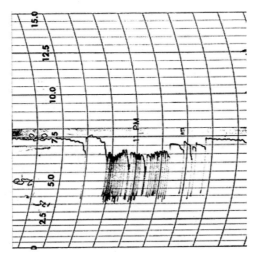

Figure 6.39   Motor current fluctuations measured on chart recorder.

C A S E   S T U D Y

A large petrochemical firm had a fan operating on an open-platform five-story tower. The fan was located in a corner of the top floor. The technician had to climb between the fan pedestal and the railing to take readings. The overhung fan was direct driven at 1,800 rpm. The technician started taking readings, and then the vibration levels jumped instantaneously to a severe range. The technician took several readings and then bailed! The vibration levels had jumped from 0.1 inch/s velocity to more than 1.0 inch/s velocity with the predominant energy all at 1× operating speed.

All attempts to balance the fan failed. Dial indicator readings were all within acceptable limits. The fan fits were checked for looseness, but none were found. Every run reflected the same pattern: smooth levels until a minute or two passed. Then the vibration shot up!

The customer was asked for the operating curve for the fan and also what the process was. Customers are not usually willing to talk about processes because they can be proprietary. This fan was part of an air recirculation system. The air went through a catalytic bed.

For readers not familiar with an operating curve, it consists of data provided by the manufacturer of the equipment that instruct the end user how that piece of equipment was designed to work. The fan datasheet indicated an optimal change in pressure from suction to discharge of 50 psi. A simple manometer was used to test the fan. The ΔP across the fan was next to *zero*! There was no backpressure against the fan at the discharge. As a result, the fan blade was going into an aerodynamic whip due to a stall.

The customer had recently redesigned the fan ducting and removed a damper on the discharge side. The catalyst had been replaced with new catalyst, resulting in almost no backpressure. The recommendation to the customer was to reinstall the discharge damper. The damper would be required until the catalyst became "dirty," at which point the damper could be opened for more flow to keep the fan on the proper area of the operating curve.

CASE STUDY

A similar type of problem involved a compressor at a chemical facility. The unit was aligned and started. Vibration readings were acceptable. The technician got a call at midnight to come back. The unit sounded normal as the technician was setting up his equipment. He asked the operator why he had called. The operator replied, "Wait, you'll see why." And at that moment the vibration jumped to a severe level. After 6 seconds it dropped back to normal. The operator said, "Now, watch. Thirty seconds from now, it will do it again!" Sure enough, 30 seconds later a repeat!

The technician requested the compressor performance curve. This was another case of a machine operating within a recirculation loop. There was a letter from the manufacturer warning the end user that the compressor was operating too far out on the performance curve and that unless a restriction was placed in the discharge side of the compressor, there was a danger of the compressor entering a stall with a resulting aerodynamic whip of the impeller.

Unfortunately, this customer did not want to be told what to do. The company paid for the compressor foundation to be epoxy filled. Having never gone back, the assumption is that the same phenomena occurred, but with lower vibration levels. This meant that the machinery bearings would now take the brunt of the severe vibration excursions, resulting in a less reliable machine.

A word about aligning electric motors. Vibration readings following alignment often identify twice line frequency vibration. This is usually caused by an uneven foot support causing a twisting of the motor frame. The result is a discontinuity of the motor stator support in the frame. The predominant energy in the motor stator is twice line frequency. Loosening one motor foot at a time while watching the live spectrum will uncover the problem foot. *Do not* loosen more than one motor foot at a time. Loosen one foot, check the vibration level, and then retighten that foot before moving on to check the next foot bolt. Once the problem foot is discovered, shut the machine down and correct the shimming of that foot.

REVIEW QUESTIONS

1. A technician is performing an impact analysis with a portable data collector. No matter how hard he strikes the object with the proper impact tool, the data are not identifying any clearly defined natural frequencies. The amplitudes of the data are all very low. What is the most common problem for this result?

2. What are the two most important factors when preparing to impact a machine or part for natural frequency identification?

3. An impact test on a vertical motor and pump head identifies two separate, cleanly defined natural frequencies (see below), one lower than the other. What is the most logical reason for this difference?

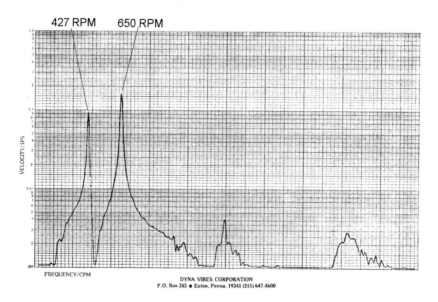

DYNA VIBES CORPORATION
P.O. Box 383 ● Exton, Penna. 19341 (215) 647-8600

4. The tall vertical pump example was successfully brought down to acceptable vibration levels by placing a long 2- × 4-inch beam between the top of the motor and a cinder block wall. What else could have been done to achieve this result?

5. The coast-down data recorded below were taken with the motor operating at 892 rpm and the pump operating at 862 rpm via an eddy-current drive, and the peak average was continued until the pump reached a speed of 622 rpm. What can the reader glean from the vibration levels that were recorded through this process?

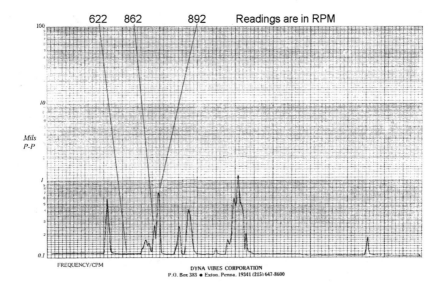

6. True or False: A successful impact analysis of a motor end bracket can be accomplished with the rotor at rest in sleeve bearings.

7. True or False: The rotor can be turning very slowly in sleeve bearings for an accurate motor end bracket impact analysis.

8. A motor, operating on 60 Hz power, has high vibration in the axial direction on the end brackets. The vibration frequency is 7,200 cpm. The motor is only running at 1,180 rpm. What is the excitation force that is exciting the end bracket?

9. The case history in this section of the book discussed a large two-pole motor, operating on a 60 Hz-power, sleeve-bearing machine with high axial vibration at 7,200 cpm. The motor was 2,500 hp, and it was coupled through a fluid coupling, with ball bearings, for speed control, to a boiler feed pump. The motor axial vibration, taken on the motor end bracket housing, was 0.6 inch/s velocity. The axial on the inboard side of the fluid drive was 0.04 inches/s velocity! Explain how that much vibration in the axial direction on the motor could all but disappear on the fluid drive inboard axial.

10. Question 9 indicated high vibration on the sleeve bearing and bracket in the axial vibration. What damage would this have caused?

11. End bracket resonances are subject to changes in support stiffness. What could change the total equivalent support stiffness on a large-horsepower motor that is bolted to a steel sole plate imbedded into a large block of reinforced concrete sitting outside on a thick concrete slab?

12. True or False: Unbalance due to a bent shaft can always be corrected by hanging a sufficient amount of weight to overcome the unbalance.

13. True or False: A worn sleeve bearing will usually be indicated by harmonics of a fundamental frequency that is less than half the shaft speed.

14. A brand-new sleeve bearing has been installed in a machine. The vibration levels on restart of the machine are not within tolerance, and the predominant peaks are at exactly one-half the operating speed and harmonics of it. What is the diagnosis?

15. A sleeve bearing machine on a closed loop oil pumped lubrication system has bearing clearances that are close to specification, yet there is a vibration at exactly 1/2 operating speed. What can be tried to alleviate the vibration?

16. A sleeve-bearing machine is running up to speed, and as it passes through a known critical frequency, vibration at the critical frequency starts to occur, and even though the machine continues to speed up, the vibration at the critical frequency continues, and it is severe. What is the diagnosis?

17. True or False: Brand-new ball bearings are being fitted to an existing pump shaft. The shaft is measured and found to be 0.002 inch over specification. Because we are looking to shrink the bearing onto the shaft anyway, this is okay.

18. A reconditioned motor with new ball bearings is placed on a truck for delivery. The shaft turns freely, indicating that the bearings are properly installed. The motor gets to the customer, and after three months of running, the bearings are showing "bad" signatures. What is the logical diagnosis?

19. True or False: A bearing that shows signs of brinelling but no sign of surface deformation or corrosion can yield a full life.

20. A motor has large rotors mounted onto its dual output shafts. A design engineer has calculated the static rotor weight to be within the specifications for bearing loading. The motor ships, and on arrival, testing proves that the bearings are cracked. What happened?

21. What is the calculation for coupling hub shrink fit?

22. Large rotating machinery has historically been fitted with sleeve bearings. Manufacturers have switched to ball bearings to lower cost. The clearances in these bearings can be so large that the balls have a problem tracking, resulting in unstable bearing operation. What is the fix for this phenomenon?

23. What is the typical number of harmonics of running speed in a spectrum taken on a machine operating within satisfactory limits?

24. A technician is laser aligning a motor-generator set. The first set of readings calls for 0.025 inch under all four feet. The next set of readings calls for 0.050-inch shims to be added to all four feet, and another reading is taken. The laser calculation now calls for 0.100 inch under all four feet. The additional shims are added, and a new set of readings is taken. Now the laser calculation calls for 0.200 inch under all four feet. What is the problem?

25. True or False: When checking for soft foot on a large, heavy machine resting on I-beam rails, it is always advisable to use the one-foot-at-a-time method called for by the laser equipment manual.

26. Alignment of a motor with sleeve bearings is critical to long life. The following true/false questions are all related to the alignment of a sleeve bearing motor:

    a. True or False: An electric motor shaft will operate on what is known as its *magnetic center*. This can be marked while the motor is under unloaded testing.

    b. True or False: It is important to scribe the shaft for the total mechanical endplay of the motor shaft.

    c. True or False: Some motors are equipped with shims that can be moved to either side of the sleeve bearing in order to achieve proper axial mechanical clearance.

    d. True or False: The motor is aligned, and a decision must be made as to whether it is better for the motor to be centered between the mechanical end stops or moved very close to one end stop to be on magnetic center. Magnetic center is more important than clearance from the end stops.

27. What is required when necking down foot bolts to eliminate a motor foot that is bolt bound?

28. A technician has just taken their first set of readings on an alignment job. The solution calls for the removal of 0.100 inch from the back feet of the motor. There are no shims under the motor back feet. The technician reverses the laser head and prism to see if the alignment can be accomplished by moving the driven equipment. Again, the solution calls for the removal of 0.100 inch from the back feet, and there are no shims under the back feet of the driven equipment. What is the diagnosis?

29. What is the radial alignment adjustment for a machine train with a 3-ft spool piece as part of the coupling?

30. True or False: Thermal alignment numbers are always available from the manufacturer.

31. True or False: The technician should never need to check exhaust piping, equipment supports with torsional pivots, and so on, because shaft alignment is all that is necessary.

32. True or False: Belt tension should be very tight. The tighter the better.

CHAPTER 7

Advanced Diagnostics

UNDERSTANDING RIGID-BODY MODES

This chapter's main intent is to educate the reader on the topic of the difference between rigid body modes and bending modes in rotating machinery. Anyone who has stood next to a machine as it comes up to speed has "felt" the vibration get worse at one frequency, only to have the vibration dissipate as the machine continues to speed up. Knowing what that is, and the causes, are essential to becoming a compete, well-rounded field service test technician.

If you're responsible for maintaining motors, you may have sent units off to be reconditioned, only to see them demonstrate high vibration levels once they're back on their foundation—and uncoupled to boot! So now what? Your vibration technicians may be reporting high 1× vibration levels, which may point back to the repair facility. If it has documented its work, the repair facility will have data from the final test stand run indicating that vibration levels were well within satisfactory limits. Both answers can't be correct. Or can they? Rather than waste time arguing with the repair shop, I suggest a more efficient approach: Take more data. Chances are that there's critical information being overlooked, and such data usually can expose your problem. If your vibration technicians don't know the answer, then they're not asking the right questions—the kind that can be answered with more data or by taking a closer look at the motor's mounting and foundation.

The discussion on alignment suggested using no more than four or five total shims under any one foot. Figure 7.1 is an example of shimming that will cause problems.

Figure 7.1 Loss of support stiffness due to excessive shimming.
(Courtesy of Ludeca, Inc.)

Vibration data that include a peak-averaged coast-down spectrum will help isolate your problem. These readings will clearly identify any areas where the resulting high vibration is due to resonance (resonance is the condition where a natural frequency is close to—and thereby excited by—a forcing function such as 1× operating speed due to residual imbalance). Impact data, when properly collected, also can identify natural frequencies. You must be careful to impart enough energy in the frequency range of interest using a soft-tipped hammer with a large attached mass. If sleeve bearings are present, the rotor must be at a slow roll and not at rest. Failure to do this will couple a portion of the shaft stiffness into the end brackets, which will then yield erroneous information.

Let's start with a little background on what we're measuring. Induction rotor construction isn't complicated: a solid-steel shaft, hundreds of thin laminations, and either solid or cast bars connected to end rings. Modeling this rotor to identify its natural frequencies is easy because a simplified two-dimensional (2D) finite-element program uses shaft lengths, diameters, and material properties.

Figure 7.2 displays the rotor geometry, its dimensions, and the locations along the length of diameter changes. The model is a quick check to make certain that data input is correct. Figure 7.3 shows a model of a 1,000-hp two-pole motor (the laminations, bars, and end rings are modeled as external weight and polar and transverse intertias, but not shown). The 2D model requires a support stiffness that most closely approximates the structure supporting the motor—for example, a motor bolted to a 2-inch sole plate embedded in grout that's part of a 16-inch-thick substrate. The end user, as the equipment operator, and the primary construction contractor are responsible for the equipment foundation. They must ensure that the rigid-body modes do not end up near major forcing functions, such as 1× and 2× running speed. In addition, the

ROTOR DIMENSIONS   3/18/2010   5:55:06

Location				External Weight and Stiffness			
No.	Length	Inside	Outside	Weight	Polar I	Trans I	Section I
	(In)	(In)	(In)	(Lbf)	(Lbf*In**2)	(Lbf*In**2)	(In**4)
1	.093	.313	.525	7.279E-01	1.021E+00	5.342E-01	0.000E+00
2	.093	.313	.525	0.000E+00	0.000E+00	0.000E+00	0.000E+00
3	.093	.313	.525	0.000E+00	0.000E+00	0.000E+00	0.000E+00
4	.093	.313	.525	0.000E+00	0.000E+00	0.000E+00	0.000E+00
5	.093	.000	.525	0.000E+00	0.000E+00	0.000E+00	0.000E+00
6	.093	.000	.525	0.000E+00	0.000E+00	0.000E+00	0.000E+00
7	.093	.000	.525	0.000E+00	0.000E+00	0.000E+00	0.000E+00
8	.093	.000	.525	0.000E+00	0.000E+00	0.000E+00	0.000E+00
9	.093	.000	.525	0.000E+00	0.000E+00	0.000E+00	0.000E+00
10	.080	.000	.438	0.000E+00	0.000E+00	0.000E+00	0.000E+00
11	.080	.000	.438	0.000E+00	0.000E+00	0.000E+00	0.000E+00
12	.235	.000	1.250	0.000E+00	0.000E+00	0.000E+00	0.000E+00
13	.235	.000	1.250	0.000E+00	0.000E+00	0.000E+00	0.000E+00
14	.235	.000	1.250	0.000E+00	0.000E+00	0.000E+00	0.000E+00
15 BRG	.235	.000	1.250	0.000E+00	0.000E+00	0.000E+00	0.000E+00

Figure 7.2   Example of a portion of data input for 2D FEAFinite Element Analysis rotor model.

floor concrete must have been poured properly, with 2 ft of crushed stone underneath, for an equivalent support stiffness of approximately 400,000 lb/inch. For whatever reason, your motor may not sit on a foundation this robust. Most industrial applications using I-beam base construction have an equivalent support stiffness of approximately 100,000 lb/inch, which is far from optimal.

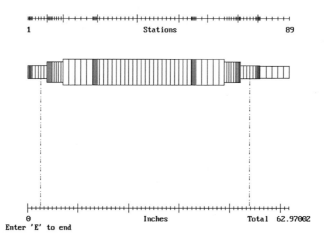

Figure 7.3   2D FEA Finite Element Analysis model of induction motor rotor.

There are other applications where the foundation has been designed specifically for lower stiffness, such as isolation spring systems. In these applications, the entire driver and driven units are isolated. This guarantees that the first two rigid-body modes are well below operating speed. However, anything that the installation contractors do to compromise this isolation will cause excessive vibration for the same reason—the stiffness has now changed, and the rigid-body mode now coincides with the operating speed. Other issues that can cause vibration include bottoming out of spring isolators or isolators that are incorrectly adjusted or improperly specified.

When I talk about lateral modes, remember that there are rigid-body modes and bending modes. The motor design engineer can predict the bending modes because they are predicated on shaft stiffness and rotor mass. Most machinery has bending modes that are higher than twice the operating speed. There are, however, machines that operate above the first and even the second bending mode (steam and gas turbines, for example). Some large motor frames operate above the first bending mode.

The end user and the primary construction contractor responsible for the foundation must ensure that the rigid-body modes do not end up near those major forcing functions, such as 1× and 2× operating speed.

Figures 7.4 and 7.5 show the rigid-body modes, with no bending in the shaft (you can see from the mode response that the bearings will be in phase for the first mode and out of phase for the second).

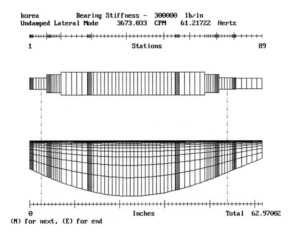

Figure 7.4   First rigid-body mode.

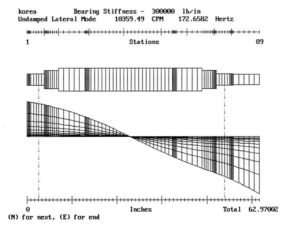

Figure 7.5   Second rigid-body mode.

Figure 7.6 shows the first bending mode. You can see that its frequency is 21,556 cpm, or more than 3× operating speed for this two-pole motor. You can also see that the mode shape curve crosses the centerline, indicating true bending. The displacement at the bearings is in phase for this mode.

Note: This is the first mode discussed on this topic that the motor manufacturer is 100% responsible for because it is determined by rotor stiffness and mass. The support stiffness has no effect on this mode.

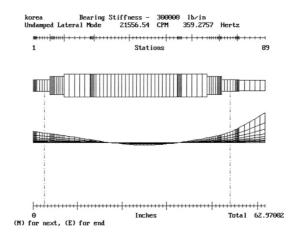

Figure 7.6   First rotor bending mode.

The critical speed map results in Figure 7.7 are the key to understanding what's happening with the reconditioned motor, now vibrating on its foundation. Note that a base stiffness somewhere between 50,000 and 100,000 lb/inch is ideal for this motor. If your foundation is such that the equivalent support stiffness is close to 30,000 lb/ inch, then the second rigid-body mode of the motor would be excited. Note the intersection of the 3,600-rpm line with the second rigid-body-mode curve and the 15,000 lb/inch equivalent support stiffness vertical line. Equivalent support stiffness is the total stiffness in series from the bearing down to the Earth. The weakest link in that stiffness chain largely determines the equivalent support stiffness:

Total stiffness 1/k total = 1/k bearing + 1/k bracket
+ 1/k foot shims + 1/k base + 1/k concrete floor + 1/k dirt

For example, if the feet have excessive shims beneath them, that stiffness will drop significantly, affecting the total support stiffness significantly.

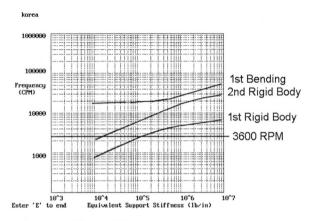

Figure 7.7   Critical-speed map.

This example also illustrates that no matter what you change in a typical support stiffness setting, the first bending mode will be unchanged. The National Electrical Manufacturers Association (NEMA) standard for testing motors on isolation pads is based on understanding the critical-speed map. The isolation pads lower the equivalent support stiffness to the 1,000–10,000 lb/inch range, moving both rigid-body modes well below the running speed. The deflection of the isolation pad must be to specification—too soft or too hard and your motor could be in a rigid-body mode on a test stand with very high vibration results. This often happens with large two-pole sleeve-bearing motors. The specification calls for approximately 10% deflection of the isolation pads due to motor weight, which must be verified to prevent excessive vibration on the test stand.

CASE STUDY

A review of the peak hold–averaged spectrum will identify natural frequencies that might be close to the operating speed. Figure 7.8 is an example of a coast-down on a large two-pole motor in all three axes on both bearings. The motor was operating at 3,585 rpm when the data averaging began, and technicians then cut the power to the motor. The processor saved every amplitude data value during the coast-down until the process was stopped at approximately 20 Hz, or 1,200 rpm. Note that the amplitudes in three of the locations increase as the speed dips below 60 Hz. This identifies the residual imbalance exciting a natural frequency and driving it into resonance. The rule of thumb for natural frequencies is to ensure that they are more than 15% removed from any major excitation force. This example encountered a natural frequency at approximately 57 Hz. This is only 5% removed from the predominant vibration source, 1× running speed.

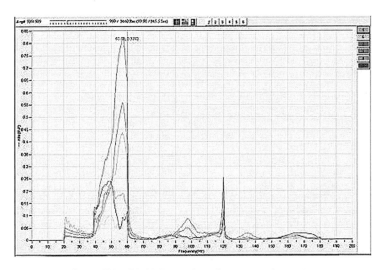

Figure 7.8   Multichannel peak-averaged spectra.

Figure 7.9 is an example taken on a large fan in the HVAC system of a 55-story building (20% of the building would be affected by the loss of this fan). The fan operates over a speed range as low as 200 rpm to as high as 405 rpm. The peak-averaged plot in the figure captured the vibration over the entire operating range, revealing a resonance condition in the fan bearing pedestal near 307.5 rpm. These fans were originally fixed-speed units. With the addition of variable-frequency drives, it became apparent that there were speed ranges that had to be avoided. This unit now has a speed range that is blocked out to eliminate staying at the problem speed. This application did not experience any damage to the fan bearings because they were sleeve bearings and the problem vibration was in the axial direction of the fan pedestal.

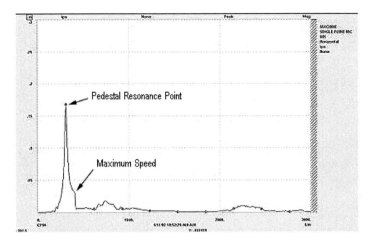

Figure 7.9   Peak-averaged plot of fan pedestal axial vibration.

How do we measure support stiffness? The measurement requires a two-channel digital signal processor that has the capability to calculate the frequency-response function (the ratio of the response motion to the input force excitation). The data are captured as calibrated impact data. Most digital signal processors will save all the data necessary to give the results plotted in Figures 7.10 and 7.11.

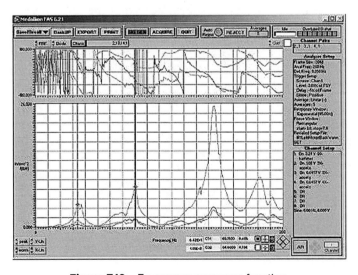

Figure 7.10   Frequency-response function.

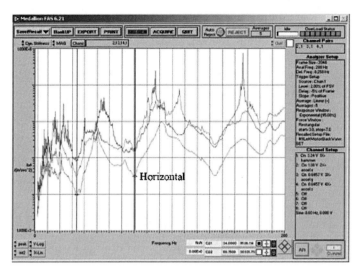

Figure 7.11   Same data used to calculate apparent stiffness.

The measurement in Figure 7.11 was taken on a large direct-current (DC) motor with a soft support system. The first cursor has been placed on the running speed of the motor (34 Hz, or 2,040 rpm). Figure 7.10 identifies a natural frequency response at 34 Hz. Figure 7.11 is the dynamic stiffness plot of the same impact test. Most dual-channel digital signal processors calculate dynamic stiffness by taking the frequency-response function (FRF), double integrating the response, and then taking the reciprocal.

Most FRF data are taken in accelerance (response acceleration in proportion to the impact energy). The double-integrated reciprocal of acceleration is apparent stiffness. Remember, it's important to keep the transducer units in order during this process. In this example, acceleration units of inches per second per second yield stiffness in pounds per inch.

Note that the horizontal trace dips at the running-speed peak. This reflects the reduction in stiffness, thereby resulting in an increase in the vibration level. The response in the FRF will be a maximum peak, and a valley will appear in the dynamic stiffness plot at the same frequency. The stiffness at this natural frequency is 9,536 lb/ft (in/s^2).

One more example:

CASE STUDY

A customer recently sent a 2,500-hp two-pole sleeve-bearing induction motor to a shop for reconditioning. Following the reconditioning work, repair technicians ran the motor at full voltage on their test stand. The test-stand foundation consisted of a 6-inch-thick steel base imbedded in a 16-inch concrete floor. The motor was set directly on the steel base with toe clamps grabbing the motor feet and securing them to predrilled holes in the steel base. The measured vibration levels on this motor were all well within satisfactory limits, and bearing temperatures also stabilized well below specified limits.

The shop sent the motor back to the customer, but shortly thereafter, the shop received a call from the customer stating that the vibration levels were not within tolerance (the motor was uncoupled at the time). Attempts to review a peak hold–averaged plot or a properly performed impact spectrum from plant personnel were unsuccessful. The motor was returned to the motor shop as defective.

Testing at the shop again revealed that the motor was good. This time, the shop decided to send a field service technician to the customer along with the motor. After several hours on site, it became apparent what was wrong. The motor feet ran the entire length of the motor housing, front to back. The customer had placed shims only at the four corners. The field service technician shimmed the center of each leg, and motor vibration dropped to satisfactory limits. Costly? Absolutely. Avoidable? Yes! A learning experience? Hopefully.

The next time you encounter a reconditioned motor that demonstrates more vibration than you'd like to see—and you've eliminated unbalance, misalignment, soft foot, and every other possibility—look at the support stiffness. What's changed? It could be something as simple as shorter shims instead of shims designed to support the entire motor foot. It could be corroded shims that don't supply full surface support when restacked. Foundations don't last forever. You may want to check for cracks in your concrete foundation or vibration in concrete pads because these would signify that the fill below the concrete has washed away. The point is that you need a complete evaluation of the operating environment, which means taking more data. If the motor had no problems when it came out but does after reinstallation, it may not be the reconditioning work that's at fault.

CASE STUDY

A reconditioned motor was returned to a petrochemical plant. The customer complained of excessive vibration on soloing the motor prior to alignment. A visit to the site revealed a substantial motor base support, yet the vibration was excessive. The customer shared that it had regrouted the entire motor base. This process had moved a rigid-body mode in the motor right up to running speed. Unfortunately, there was no way to modify the base. It was as stiff as it was going to get. The motor was in a constant-speed application. The decision was made by the contractor to design and build two dynamic absorbers to reduce the vibration at running speed. A *dynamic absorber* is a simple spring-mass system designed to vibrate at the natural frequency of interest. The natural-frequency response of the machine is split into two new natural frequencies: one lower and one higher than running speed. There are many formulas for designing dynamic absorbers. I am offering a spreadsheet designed by "Electric Pete," better known outside the vibration blog world as Peter Schimpf. He is a reliability engineer for a nuclear power facility and has always gone the extra mile in helping others in the vibration and electrical engineering world. The spreadsheet is offered with Pete's permission.

Variable Description	Symbol	Value	Units	Formula	Comments
Inputs Bar Cross Section (select circular or rect)	Xsect	circular		For list-of-values, click on cell, then click on down-arrow	just to right of cell
Length (English units)	Le_	20	inch		Input
Distance to weight (English units)	ae_	18	inch		Input
bar diameter (circ Xsection only) - engl units	De_	1.5	inch		Input
bar width (rect Xsection only) - engl units	be_		inch		Input
bar height (rect Xsection only) - engl units	he_		inch		Input
Density (English Units)	rhoe	0.282	lbm/inch^3		~0.282 typical for steel
Young's Modulus (English units)	Ee	2.90E+07	lbf/inch^2		~3E7 typical for steel
Target resonant frequency (English units)	Fncpm	3600	cpm		Input
Intermediate Variables Length (SI units)	Ls_	0.5080	m	=Le_/39.37007874	Unit Conversion
Distance to weight (si units)	as_	0.4572	m	=ae_/39.37007874	Unit Conversion
bar diameter (si units)	Ds_	0.0381	m	=De_/39.37007874	Unit Conversion
bar width (S.I. Units)	bs_		m	=be_/39.37007874	Unit Conversion
bar height (S.I. Units)	hs_		m	=he_/39.37007874	Unit Conversion
Density (S.I. Units)	rhos	7805.73	kg/m^3	=rhoe*27679.90471	http://www.onlineconversion.com/density_common.htm
Young's Modulus (S.I. Units)	Es_	1.99948E+11	N/m^2	=+Ee*6894.757293178	http://www.onlineconversion.com/pressure.htm
Area Moment of Inertia (si)	Is_	1.03436E-07	m^4	=IF(Xsect="rectangular",bs_*hs_^3/12,PI()*Ds_^4/64)	I = b * h^3 / 12 or PI* D^4/64
Distributed Mass Per bar Length (S.I.)	m1s_	8.899	kg/m	=rhos*IF(Xsect="rectangular",hs_*bs_,PI()*Ds_^2/4)	(mass/meter) = (mass/vol) * Area
Target Natural Frequency (S.I. Units)	Fns	60	hz	=Fncpm/60	Unit Conversion
Target radian natural frequency (si)	wns	376.991	rad/sec		Unit Conversion
Outputs DUNKERLY - M2 concentrated Mass (si)	M2ds	3.0677	kg	=3*Es_*Is_/(as_^3*wns^2)-0.2419355*m1s_*Ls_^4/as_^3	See tab entitled "Derivation"
DUNKERLY - M2 concentrated mass (engl)	M2de	6.7631	lbm	=M2ds*2.204622622	Unit Conversion

Figure 7.12a   Excel spreadsheet for design of dynamic absorber.
(Courtesy of Pete Schimpf.)

The piece of metal that supports the mass should not be round or square but have a rectangular cross section. This is to maximize the response in the direction with the highest resonance response. Also, the sizing of the spring should ensure several years of operation before fatigue sets in. Remember, it's now going to be vibrating the entire time the motor is in operation. The mass also should be approximately 3% of the main mass: 2,000-lb motor × 0.03% = 60 lb, or 30 lb per end.

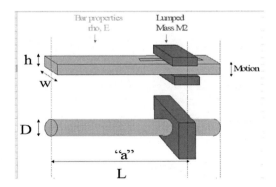

Figure 7.12b   Dimensions of inertial mass examples.

Derivation of formula used to find M2 (Dunkerely approach)

> # ==== PART 1 - SYMBOLS ======
    E = Young's Modulus
    I = area moment of
    inertia
    m1 = distributed mass per length of the bar
    M2 = discrete mass attached distance "a" from fixed end
    a = distance mentioned above
    L = length of
    bar
    w1 = radian nat freq of system with mass m1 only (no M2)
    w2 = radian nat freq of system with mass M2 only (no m1)
    wn = calculated radian natural frequency

==== PART 2 - CALCULATIONS ====
    1/wn^=1/w1^2+1/w2^2   # Dunkerley's approximation
                                         # Den Hartog App 5
    w1^2:=12.4*EI/(m1*L^4)   (12.4=3.52^2)
    w2^2:=3*EI/(M2*a^3);   # Den Hartog App 1, eqn 2

    Solve for M2:
    M2 := 3*EI/(a^3*wn^2)-.2419354839*m1*L^4/a^3

You can create the same natural frequency with a thin (flexible) bar and small mass or a thick bar and large mass.
    Thicker bar and larger mass are less susceptible to fatigue and provides larger separation of amplification
    "sidebands" around the resonant frequency
    Some recommend absorber mass 3% of main mass

L is not really the length of the bar. It is the effective length to the point where the fixed boundary condition exists
For example, a 30" bar clamped or welded for the first 2" of its length should use
L=28"
The distance should also be measured from the same point

Provide provisions for adjustment to tune the absorber
See Randy Fox' paper and maintenanceforums.com for more application suggestions
    For example:
            http://maintenanceforums.com/eve/forums/a/tpc/f/3751089011/m/3211047223

This spreadsheet prepared by
electricpete

Test Case Problem
Parameters
    Target Frequency: Fnat = 1800 cpm (30HZ)
    Rectangular Bar
    b=2"   L=24"
    h=1"   a = varies between 16 to 24" at right->

Steel (rho=0.282 lbm/inch^3, E=2.9e7 PSI)

a (inch)	Computed W2 (lbm) Method: Fox	Computed W2 (lbm) Method: Dunkerly	FE calculation of Fnat (HZ) * Using Fox W2	FE calculation of Fnat (HZ) * Using Dunkerly W2
24	6.314	8.123	32.80	30.08
22	7.214	10.546	34.11	30.02
20	8.393	14.036	35.51	30.00
18	9.978	19.254	37.02	30.00
16	12.177	27.415	38.67	30.05

* - Calculated using transfer matrix program - available on request
Based on Euler Bernoulli Beam Model
Shear Deformation neglected
"Rotating Inertia" (disk effects) neglected

Conclusion
    Dunkerley creates the target frequency (30hz) much better than Fox' equation
    (Of course, accuracy is not critical if sufficient room for adjustment is provided)

Figure 7.12c   Terminology for spreadsheet in Figures 7.12a–Figure 7.12c. Comparison of Fox and Dunkerley Methods for Calculating Dynamic Tuned Absorbers.

Let's consider a multi-shafted rotor model.

C A S E   S T U D Y

A large paper manufacturer was experiencing broken motor shafts. This model was completed to diagnose the failure of several motor shafts. The shaft was fatiguing and breaking off just inboard of the coupling to the jack shaft. The motor was on a variable-frequency drive. The roll was part of the rewinder process. Figure 7.13 indicates that the first rigid-body mode of the paper roll was 1,449 rpm.

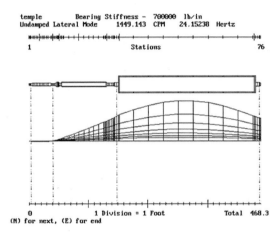

temple   Bearing Stiffness - 700000 lb/in
Undamped Lateral Mode   1449.143 CPM   24.15238 Hertz

1   Stations   76

0   1 Division = 1 Foot   Total   468.3
(N) for next, (E) for end

Figure 7.13   Example of 2D finite-element analysis
model of motor, jack shaft, and large paper roll.

Multichannel vibration data were recorded on the rewinder during its entire operating cycle. It was discovered that the motor, which was variable-frequency driven, would start slowly and then go through several acceleration ramps until achieving the maximum speed for operation. A vibration frequency was identified that was predominant during these fast acceleration ramps (Figure 7.14). The rotor model identified a torsional mode shape that had a high degree of torsional bending (twisting) in this frequency range (Figure 7.15).

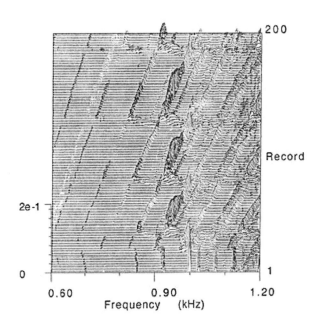

Figure 7.14   Acceleration ramp from variable-frequency drive
exciting a natural frequency near 900 Hz.

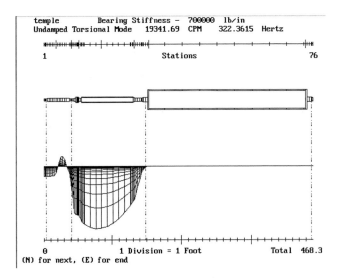

Figure 7.15   Torsional mode.

The slope of the line is the key to viewing a torsional-mode shape. The sharper the slope, the greater the torsional bending. This mode identified a high torsional bend in the area of the output shaft of the motor. The motor shaft was found to be slightly undersized compared with the other rewind motors. A motor with a larger output shaft diameter was chosen, and the problem was solved.

Another fix for this problem would have been to reduce the attack angle of the acceleration ramp when the drive made a speed increase. This also would have solved the breakage problem.

A processing plant had four large DC motor–driven fans on an upper floor of the plant. The motors had experienced numerous armature and bearing failures, but the fan bearings had good reliability. Vibration readings were recorded on one of the fans (Figures 7.16–7.18).

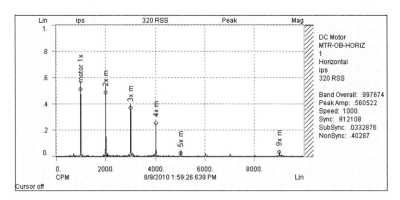

Figure 7.16   High 1× vibration on motor outboard horizontal.

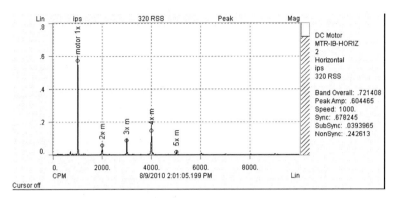

Figure 7.17 High 1× vibration on motor inboard horizontal.

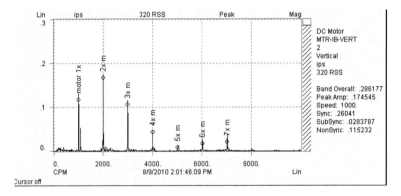

Figure 7.18 Motor inboard vertical.

The ratio of the 1× vibration between the horizontal and vertical readings is more than 5:1. This is indicative of a natural-frequency issue. The motors were mounted on very large concrete columns, so a rigid-body mode of the motor armature was suspected. A rotor model of the motor was completed. The model included the motor rotor, coupling, and fan.

The critical speed map indicated that the first mode crossed the speed frequency (1,000 rpm) at approximately 50,000 lb/inch. This is extremely low for large industrial equipment (Figure 7.19).

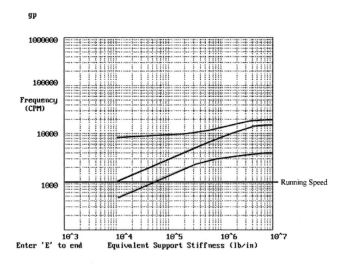

Figure 7.19 Critical speed map.

The motor was obviously resting on something that lowered the equivalent support stiffness so low that the first rigid-body mode was close to operating speed (Figure 7.20). The first bending mode was more than 10 times the operating speed (Figure 7.21). The motor engineers had done their job.

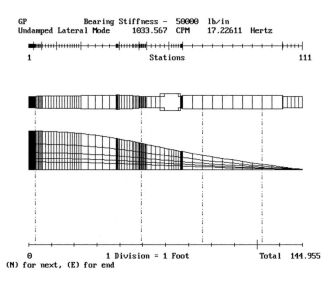

Figure 7.20 Mode shape of first rigid-body mode.

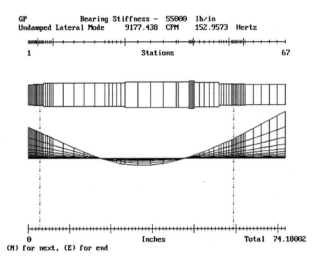

```
GP              Bearing Stiffness -  55000  lb/in
Undamped Lateral Mode      9177.438  CPM    152.9573  Hertz
```

Figure 7.21   First bending mode.

The decision was made to perform an ODS Operating Deflection Shape on the entire system. Triaxial accelerometers were moved from point to point until all 53 locations were recorded. The 1× running-speed mode deflection identified a large movement in the floor below the concrete pedestals. An inspection of the floor from one story down found large 36-inch web I-beams supporting a corrugated steel floor. There was no connection from the floor to the I-beams. The concrete floor was a 3-inch pour on top of the steel. This was the point where the necessary stiffness was lacking (Figure 7.22).

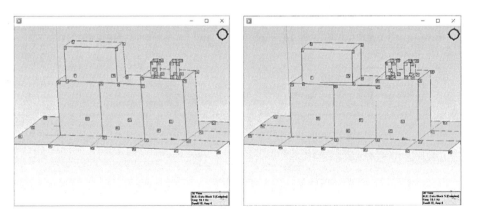

Figure 7.22   Operating deflection mode shape of the motor-fan system.

The recommendations to the customer called for steel plates to be placed below the I-beams and on the concrete deck next to the concrete pedestals. Then long threaded studs would be connected to the plates through drilled holes to anchor the "floating" concrete floor to the beams. This would raise the equivalent support stiffness enough to move the rigid-body mode away from running-speed range. The customer to date has completed this modification on one fan with excellent results. All vibration levels have dropped to satisfactory levels.

EXAMPLE OF BEST OEM PRACTICES COMPROMISED

Isolation pads are used for many applications, including alleviating vibration in an inhabited building. They are also used to lower rigid-body modes of rotating machinery away from excitation forces. The OEM shipped the materials with instructions for use to prevent a potential vibration issue on site installation of its machine. The example in Figure 7.23 shows how *not* to install such a machine. If multiple layers are used, the cross-ridge pattern of the isopad requires a thin metal sheet to distribute the load. These pads have been crushed. The end-user's contractor also added hold-down bolts that were cocked into the base. This also created a contact point between the base and the bolt that compromised the isolation desired result.

Figure 7.23   How *not* to install isolation pads.

Vibration levels on this unit exceeded 1 inch/s velocity because of the rigid-body mode coinciding with 1× running speed (Figure 7.24). This unit ran at the factory with vibration levels under 0.1 inch/s.

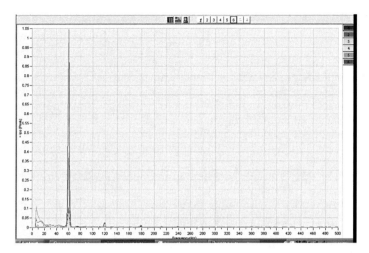

Figure 7.24 Loaded chiller vibration.

The radial vibration of a machine can help identify the difference between unbalance and a suspected resonance due to either a localized natural frequency or encountering a rigid-body mode. Unbalance should result in the 1× vibration magnitude being similar in both the horizontal and vertical directions. The difference between the two readings will indicate a difference in stiffness and damping. Typically, the ratio should be approximately 2:1. The preceding example had a 10:1 ratio between the vertical and horizontal directions. This is obviously not the result of unbalance, bent shaft, or misalignment.

The phase relationship of the two ends of the motor on this compressor indicate a 180-degree phase difference. The isolation pads were engineered to lower both rigid-body modes to be lower than running speed. However, because of poor installation, the second rigid-body mode now resides close to the 1× running speed (Figure 7.25).

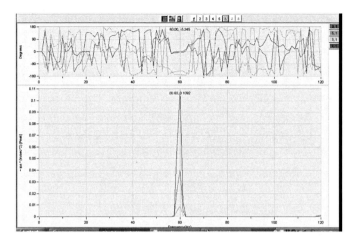

Figure 7.25   Magnitude and phase of 1× vibration.

A unit at the same location had results that clearly show the result when a unit has been isolated successfully. The vibration levels were 75% *below* the OEM specifications.

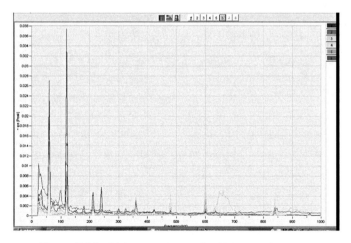

Figure 7.26   Unit operating well within specifications.

The investigation of the isolation of a machine should not end with foot support. Any connection to the machine may not have been installed properly. This includes piping (isolation couplings) and power conduit. Piping also should have enough support to the building and not place any piping weight on the machine.

ENGINEER OR SCIENTIST?

C A S E   S T U D Y

I was asked by a large OEM to commission a motor at a power plant. The motor was 12,000 hp. The contact informed the author, "You are here to take readings on 2B. That's because 2A is no longer with us." Strange comment, further explanation required! "Someone working for the fan manufacturer wanted to make the fans more efficient, so he fabricated conically shaped steel pieces to 'curve' the incoming air into the fan inlet cone. Come on, I'll show you."

We walked over to a warehouse where the components had been laid out. The cones had been manufactured in two pieces to clamp around the shaft. Each half weighed 75 lb. The halves had been bolted together with four ¼- × 20-inch bolts! The customer contact continued to describe the horrible outcome:

> The fan was ready for its first operation. All the plant personnel turned out to watch. The motor sits on a 15-foot-high concrete pedestal. The "brass" were standing at the base. The fan started and came up to speed, 900 rpm. The fan sounded fine. Then the four pieces of steel decided to come loose! Imagine 300 lb of "loose" steel. The motor pulled every 2-inch base bolt out from the sole plate and came up off the base. The top hat (2,000-lb motor cooling hood) flew off the motor as manmade lightning issued forth from the motor frame. The back-motor bearing blew out 6 inches, hanging from the motor shaft. The fan walked 15 ft out of the fan housing in the axial direction! The top hat came down and caught on the railing at the top of the concrete pedestal. If it had cleared the railing, the spectators would have been crushed.

The contact then walked me over to the fan. The entire fan was surrounded by scaffolding with cutting torches ablaze everywhere. The entire fan had to be removed before repairs could begin.

I had only one question concerning 2B: "Have the cones been removed?" The answer was "Yes."

The next day, as the locks and tags were being removed, I walked down to the fan to await the startup. Standing at the base of the motor, it occurred to me that no one was standing next to me or in front of me. I turned to look back from where I had started. There were 20 people standing behind a yellow tape approximately 150 ft away. Once bitten . . .

The fan came up to speed, readings were taken, and the motor was given a seal of approval. So scientists, if they want to get involved in machinery design, might want to have their ideas verified by an engineer. I told this story to my controller when I arrived back at the office. The controller had one comment: "Aren't those bolts too small?"

THIS CUSTOMER PAID FOR THE ENGINEERED INSTALLATION REPORT

A customer in the Midwest had a new installation that included 20 blowers, all on variable-speed drives (VFDs). As part of the package, the customer paid for the OEM engineering report for proper installation. The local sales engineer called the home office when the customer complained that 50% of the motors had blown their windings. A copy of the engineering report was sent to me. The report correctly stated that *no more than one set of motor leads should be run through any one conduit.* This is critical to eliminate cross-talk between phases that will cause large-voltage standing waves. These standing waves could cause winding failures. The sales engineer also sent a picture that he thought might help in identifying the cause of the problem (Figure 7.27). The contractor obviously did not follow the engineering installation report.

Figure 7.27   Power leads for all 20 VFD motors.

VFD specifications add another whole dimension to getting it right. A customer asked me to test a small 15-hp motor to see if the winding was good. The motor was sitting on the workbench in the customer's shop. The winding was open (bad). The customer claimed that this was the second motor to fail on the same application. The motor was on a VFD. I asked about the cable run. Not only is it important to run the cable through its own conduit, but the maximum length of the run is specified by the VFD manufacturer. The customer said that the run was approximately 300 ft. The specification for the drive was 75 ft. There are ways around this. The user would need to install what is called a *terminator* at the motor. This is a choke that eliminates any standing waves on the power cables before the power gets to the motor winding.

CASE STUDY

A customer with multiple container cranes was experiencing high vibration on the main hoist motors. The type of crane is referred to as *machinery on trolley*. The main hoist motors, gearboxes and their associated cables all travel with the trolley. This saves on cabling costs. The other type of crane has the main hoist motors mounted in the machinery house on top of the crane. The cables then run from the house to both ends of the crane boom, and have enough cable for the spreader to reach the ground. The small white rectangle just to the right of the right support leg of the crane is the trolley house (Figure 7.28). That housing moves with the spreader and contains all the equipment for the main hoist. The main hoist motors consisted of two 800-hp DC motors driving a common gearbox. The gearbox output was the two cable reels for the spreader.

Figure 7.28 Typical container crane with trolley house hoist system.

The vibration was so severe that the armature of the DC motors was impacting the stationary coils and self-destructing. The vibration was low during the entire acceleration. Then, as the speed reached maximum, the magnitude would increase exponentially. This is usually a case of a natural-frequency excitation. The manufacturer of the motors paid for a finite-element study of the entire trolley house. No natural frequencies were found near the maximum operating speed of the motors.

Data recorded on these units were extensive. The couplings were removed and precision balanced. The motors also were precision balanced when brought to the shop for repair. Testing was proposed using two proximity probes, one to read the reaction of the motor coupling hub and the other to read the gearbox coupling hub.

The equipment was installed, and proximity probes were gapped to the couplings (Figures 7.29 and 7.30). The crane operator was instructed to run the picker up from stop to full speed and then down at full speed. The analyzer was set up for peak-averaged plotting. Again, this mode saves the maximum amplitude encountered at every frequency as the running speed of the motor travels from zero up to maximum speed and then down again. The captured plot had the answer. Wherever the motors stopped accelerating and then maintained a constant speed, the 1× vibration jumped severely (Figure 7.31).

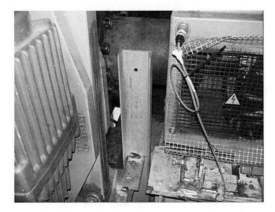

Figure 7.29   Proximity probe support channel beam for motor coupling measurements.

Figure 7.30   Proximity probe support channel beam for gearbox coupling hub measurements.

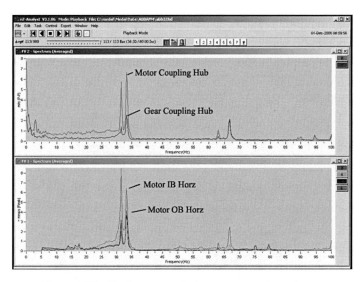

Figure 7.31   Peak-averaged plot, run up and run down.

The available AVI file has four data points, two accels on the motor horizontal locations and the two proximity probes. These are color coded. The red trace is the proximity probe reading the motor coupling. The blue trace is the proximity probe reading the gearbox coupling. The green trace in the lower graph is an accelerometer reading the motor inboard horizontal. The orange trace is an accelerometer reading the motor outboard horizontal. Notice the high vibration at the motor inboard bearing and motor coupling from the proximity probe.

The crane operator did not reach full speed on the spreader assent. He stopped accelerating with the motor at approximately 1,600 rpm. Then, on the decent, a maximum speed of 2,100 rpm was reached. Notice that the 1× vibration in both "constant speed" situations had high readings. Also notice how low the vibration levels were over the rest of the running-speed span (Figure 7.31).

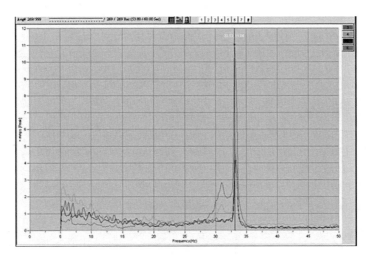

Figure 7.32   Run-up vibration on the motor and couplings.

There were two accelerometers mounted to the motor (1 inboard horizontal, 1 outboard horizontal) and two accelerometers mounted to the gearbox (1 inboard horizontal, 1 outboard horizontal). The reaction of the motor frame at maximum speed is evident. The motor accelerometers did pick up a natural frequency in the motor, just below maximum speed, but it also has a high degree of damping (energy spread over a wide frequency range).

There was something else besides a natural-frequency issue causing the problem. For some reason, the shaft centerline of the motors and gearbox were not remaining fixed. The coupling was the culprit.

Container cranes are usually specified to have Bubenzer couplings on the main hoist motors. These can absorb the shock load from an emergency stop of the spreader. A Bubenzer coupling is shown in Figure 7.33.

Figure 7.33   Bubenzer coupling.

A Bubenzer coupling has a flexible insert that looks like a circle of rubber drums connected by a rubber ring. Once the motor-gearbox has been laser aligned, there is so much clearance around the insert that the technician can move the drum part with a finger. There is no contact between the coupling jaws and the insert at rest. The torque load drops way off when a constant speed has been achieved. The stiffness of the trolley frame construction isn't enough to keep the two motors in line with the gearbox when lack of torque allows for play in the insert. This allows the couplings to move opposite each other at constant speed, creating a misalignment (Figure 7.34).

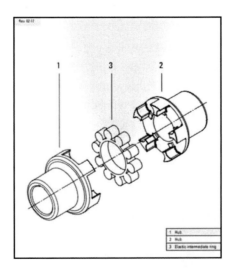

Figure 7.34   Bubenzer coupling flexible insert.
(Source: http://pintschbubenzer.de/files/bubenzer-de/downloads/epapers/couplings.html.)

Changing the coupling is not an option. Solution? The motion of the coupling at constant speed was countered by adding washers to the coupling hub connector bolts. Credit where credit is due: This solution was the result of three young engineers from the crane manufacturer working very diligently over a holiday week in New York City up in a cold crane.

Now fast-forward to a customer with the same problem but with very different equipment. The customer had 10 cranes, and all 10 had been limited to 60% rated speed because the vibration was so severe. The success in making money at an unloading facility relies 100% on the speed of the cranes. This customer had stationary house–mounted main hoist motors (Figure 7.35). Again, resonant frequencies were suspected, and the customer had tried adding mass on top of the motors (400-lb steel blocks) and welded gussets in the base below the motors without success.

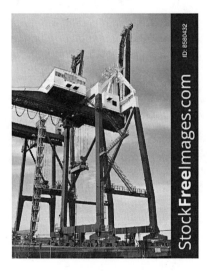

Figure 7.35   Typical stationary house for main hoist motors.

The motors were from a different manufacturer; even the gearbox manufacturer was different. The only common factor was the Bubenzer couplings. I realized, armed with my knowledge of the preceding cranes, that this job was going to yield success beyond the customer's wildest dreams. The site reliability engineer was instructed to bring a whole lot of washers that would fit the coupling bolts. By lunch, three cranes running at 100%! By the end of the week, all 10 cranes were fixed. The Bubenzer coupling is well-suited to the rest of the requirements, so a quick trim balance on the couplings was a very minor fix with a very large upside.

CASE STUDY

An aluminum mill had been experiencing repeated commutator failures on the main-drive motors for the mill. There was one 1,500-hp DC motor in tandem with two 750-hp DC motors. The customer requested an operating deflection shape of the three motors. The motors operated at 750 rpm. The problem caused sparking even with new, properly seated brushes running on a new, polished commutator.

The 1× running speed ODS did not identify any issues. However, the 2× running speed ODS identified a severe misalignment that resulted in a high 2× response in the one 750-hp motor (right) that was 180 degrees out of phase with the 2× response of the 1,500-hp motor (second from the left; Figure 7.36). The misalignment was pivoting through the other 750-hp motor (second from the right).

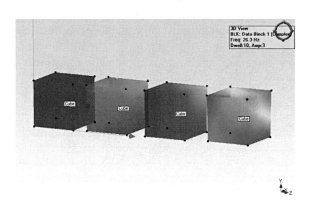

Figure 7.36   Operating deflection shape of 2× running speed.

Misalignment can cause premature bearing failure, but these motors were experiencing extreme brush wear and, as a result, commutator wear. What could have been the cause of this additional damage? The brushes and brush support arms for these motors, owing to the motor horsepower, were large. They were supported through bolted connections on the non-drive end (NDE) bracket. A modal analysis was performed on one of the brush arms. The result indicated that the first cantilever mode of the brush arm was 1,500 cpm. The misalignment vibration was exciting the brush-arm natural frequency. This motion was causing the brushes to make and break contact with the commutator surface, causing sparking and premature failure of the brushes and commutators (Figure 7.37).

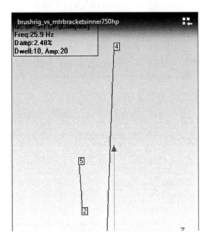
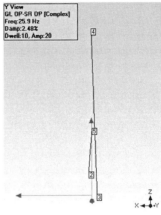

Figure 7.37 Extremes of the brush-arm motion.

This represents a simple modal analysis. Points 2 and 5 in the figure mark one of the brush support arms. Points 3 and 4 represent the commutator body. Again, as with operating deflection shapes, the movement is amplified for clarity. The movement of the brush support arm is much greater than that of the commutator body. This motion would work against the brush tension and cause sparking.

The modal analysis identified a natural frequency of the brush arm at 25.9 Hz, or 1,554 cpm. The decision was made to add a Phenolic halo support to attach to the open ends of the four brush arms. This raised the natural frequency of the brush arms significantly. The motors were also realigned. This resulted in a reduction in the vibration at 2× running speed. Notice that this modal analysis model has only four points. The model doesn't have to be complicated, just sufficient to contain the correct information.

What is the difference between modal analysis, finite-element analysis and the operating deflection shape? Modal analysis is a static test on an actual structure to identify natural frequencies of that structure and the mode shape associated with each one. Finite-element analysis is a theoretical experiment where assumptions are made concerning material properties, stiffness of connection points, loads, and so on. The geometry of the structure is created in the computer program. The software then calculates the natural frequencies and associated shapes. The loading of this model can be input to resemble the expected loads from dynamic (running) operation. The operating deflection shape is created from vibration data measured on the structure while in operation. The geometry is input by the user to approximate the structure under test and is much less complicated than finite-element analysis.

Both modal analysis and operating deflection shapes require the geometry to be robust enough to accurately capture the different motions at different frequencies. For example, if there are not sufficient points, a higher-frequency mode may resemble the same shape as a lower-frequency mode. Finite-element analysis, once given the overall dimensions of the shape of the structure, will generate the mesh automatically (Figure 7.38).

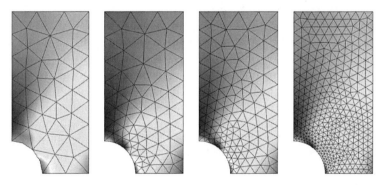

Figure 7.38   Examples of different mesh sizes
for the same part for finite-element analysis.

Modal analysis requires data collection from every point in the geometry. The I-beam section in Figure 7.39 has a total of 54 locations that require data acquisition. Data can be gathered by different methods. A modal hammer can be used for excitation, and a triaxial accelerometer can be used for response measurement. A triaxial accelerometer will require three data channels for capturing the data, but data-acquisition time will be reduced by 67%. The excitation is made by impacting the same point each time as the triaxial accelerometer is moved from point to point. The excitation only need be made in one direction because the damping within the structure will guarantee that energy is dispersed in all three directions.

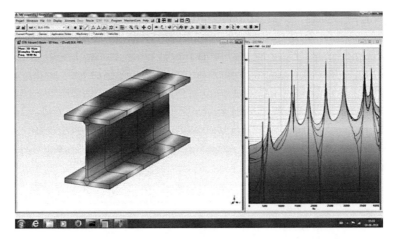

Figure 7.39   Geometry for modal analysis of I-beam section.
(Source: https://www.dewesoft.com/applications/structural-analysis.)

The modal impact hammer uses a force transducer to measure the energy imparted into the structure. The durometer of the tip and the mass of the hammer are critical to ensure the desired results. For example, the tips of impact hammers range from steel to a tip so soft it can be compressed one-third of its diameter just by squeezing it between the thumb and forefinger. The key is knowing where the frequencies are that need to be excited. For example, a large motor frame obviously should be tested to ensure there are no natural frequencies near running speed or the electrically generated frequencies. A 3-lb hammer would be sufficient to excite the motor frame. Because the frequencies of interest are low, the softest hammer tip should be used. With the soft tip, the energy will now be centered in the lower frequencies of interest. Additional mass also will help accentuate those frequencies (Figure 7.40).

Figure 7.40   Examples of modal impact hammers.
(Image courtesy of PCB Piezotronics, Inc.)

James Deel, a dear friend who taught me everything about modal analysis while I was working at Zonic Corporation in Milford, Ohio, gave me two pointers while attempting a modal analysis on a very large motor on a long winter night in northern New York State. The first was to always use an accelerometer with a high sensitivity for large structures. The second was to add mass to the excitation hammer. A hole was drilled in a 20-lb block of steel that was added to the back end of the modal hammer. The 10-mV/g accelerometer was shelved, and instead, an accelerometer with a 1.0-V/g sensitivity was used. The next impact on the structure was like someone threw a switch! All the frequencies of interest were now there. These two key elements provided a wealth of results on the very first survey. Imagine taking data on 50–100 points, curve fitting all the data, and finding the results worthless. Setting up the test equipment and deciding on what parameters are important to the test are essential to getting good results. Rest in peace, Jim.

A frequency-response function (FRF) is used to perform modal analysis on a structure. The measurement requires two inputs: a calibrated hammer or shaker and a calibrated accelerometer. The measurement, when recorded for one location on the structure, identifies natural frequencies that are responsive at that location (peaks) and natural frequencies that are at a node for that location (negative peak). Because the input and response energies are both measured, the total energy above and below unity (1.0 Accleration/Force [A/F]) is equal. The peaks represent highly responsive natural frequencies for a given location. The valleys, or *antinodes*, are natural frequencies with zero response at the present location, but they will be highly responsive at other locations—resonant points and anti-resonant points. Viewing the animated modal results makes this point quickly. There will be points in the shape with *zero* movement. This means that at that frequency and that location, the result was a valley in Figure 7.41.

A modal analysis requires the information in this figure for every point on the structure. Once all the locations have been measured, the software performs a curve fit on the data, and the resulting animated mode shapes can be viewed on the computer. These can be extremely valuable for design engineers to prevent forcing functions (normal vibration that occurs in the machine) from coinciding with the natural frequencies of the structure.

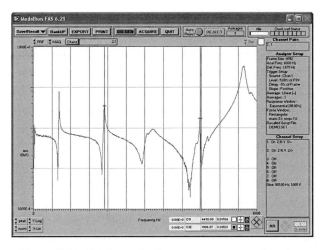

Figure 7.41 Driving point frequency-response function.

CASE STUDY

A customer had three motor cooling fans explode while in operation. Pieces of the fan penetrated the ¼-inch-thick side cover plates on the rear of the motor. A solution to this problem was urgent. A modal analysis of the fan is provided in Figure 7.42. The locations measured represent every blade connection to the backplate and the eye. Notice the blades (short connecting lines around the circumference) bending.

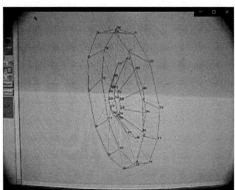

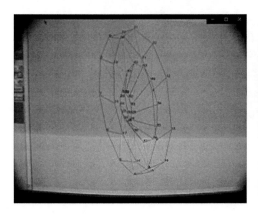

Figure 7.42 Modal analysis of motor cooling fan.

The mode was identified as 193 Hz. This is high and nowhere near 1× or 2× motor running speeds. However, the motor was driving a rotary lobed blower. These blowers are noted for having large multiples of harmonics of running speed, as well as the fundamental and harmonics of the lobe frequency. The FRF for this mode was underdamped. As a result, one of the harmonics of the blower was close enough in frequency to this mode to excite it. The mode shape indicated that the back plate and inlet eye plate were in a torsional mode, out of phase with each other.

Therefore, each torsional cycle flexed and weakened the welds of the blades to the eye and back plates. After no more than a week of operation, the cyclic bending was enough to detach the blades from the eye and back plates, and the fan would collapse axially on itself and then catastrophically fail. Imagine how proud we were at the Philadelphia Service Center when an engineer from the end user visited our facility, looked at the mode shape on the screen, and uttered, "Wow! I didn't expect to see this kind of analysis."

Interestingly, the cooling fan in this motor design was on the drive end (DE) of the motor. Placing the cooling fan on the NDE of the motor would have allowed the torsional impulses to be damped through the rotor core of the motor, possibly saving the fan. The decision was made to rebuild the fans with heavier steel stock and larger weldments. This raised the natural frequency of the torsional mode to a much higher frequency and added damping to the fan.

The complete opposite end of the spectrum would be the testing of a small cast-aluminum compressor wheel. Here the frequencies of interest are much higher, and the mass of the test piece is so light that the accelerometer used must be very small (Figure 7.43). The smaller the accelerometer, the higher the frequency range, so that works out!

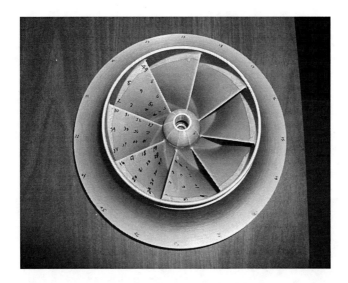

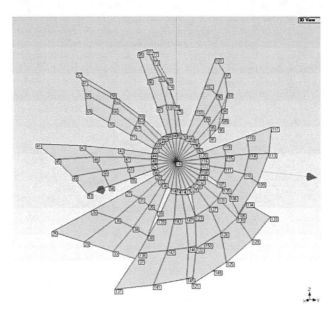

Figure 7.43 Numbered locations for modal analysis
of compressor wheel and modal geometry file.

A roving excitation point was used versus a roving measurement point for this item because the blades were so thin—even a miniature accelerometer would affect the readings. The accelerometer was mounted to a rigid point on the discharge side of the wheel. The impact hammer also must be very small because the distance between measurement locations is close.

The hammer in Figure 7.44 uses flexible plastic tubing approximately 3/16 inch in diameter. The user can position the hammer point over the test location, pull back on the head of the hammer, and let it go. The spring will allow the hammer to make a clean impact and then move off the test piece.

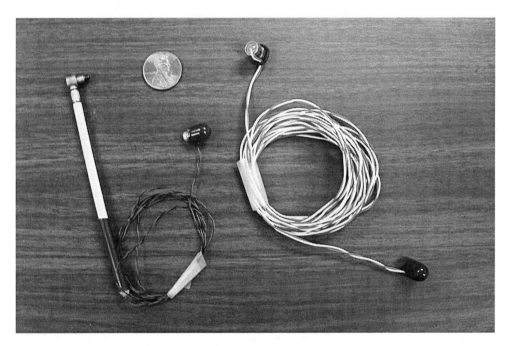

Figure 7.44   Miniature modal impact hammer.
(Image courtesy of PCB Piezotronics, Inc.)

Operating deflection shapes do not require the complex geometries necessary for finite-element analysis or modal analysis. The example in Figure 7.45 was designed to illustrate that the weakness of the motor and fan support structures can cause severe drive-belt fluctuations. This will usually prove detrimental to both belt and bearing life.

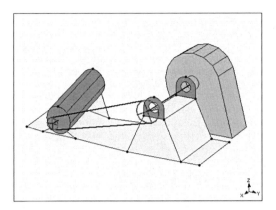

Figure 7.45   Example of simplified geometry for ODS.
(Courtesy of Acutronic.SE.)

Note: Integrated Power Services now uses RDI Technologies IRIS
M amplified motion camera. This technology has obsoleted the necessity
of collecting data, point by point, for an ODS. The technician can view
the actual amplified motion of the entire machine, structure, and so on.
The example at the URL site illustrates the results of a degraded concrete
foundation beneath the outboard fan pedestal. Thanks to Mr. Marc Amato, CEO at
Innova Logic of Saunderstown, RI for the linked example.

REVIEW QUESTIONS

1. How many shims are too many under any one foot?

2. What valuable analysis should be completed when a machine or motor under test is getting ready to be shut down?

3. True or False: The motor design engineer is not responsible for the frequencies of the rigid-body modes.

4. What is the critical speed map good for?

5. The plot below is a peak-averaged coast-down of a 2,000-hp motor. The operating speed is 3,585 rpm. What very important piece of information was captured in this plot?

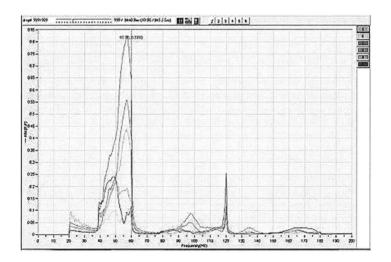

6. The plot below was recorded on the outboard fan pedestal on a variable-speed-controlled motor-gearbox-fan. As the fan slowed down, there was a 4× increase in vibration. This pedestal is situated on the eighteenth floor of an office building. There is no way to increase the stiffness of the pedestal. What should this customer do to prevent the fan from stopping at this obvious critical frequency?

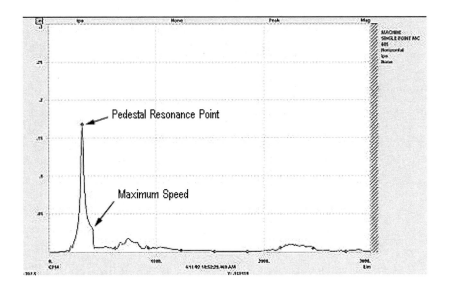

7. Below is a critical speed map of an 800-hp DC motor sitting on a 6-ft-high concrete pedestal. The motor usually operates at 1,000 rpm. The vibration levels, which were excessive, occurred with the motor horizontal ends in phase. Rigid-body modes considered, what can we say based on the results of where the equivalent support stiffness resides? (Remember, the first rigid-body mode is the lowest of the modes in frequency.)

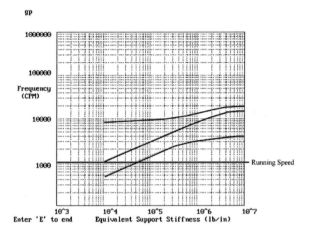

8. Below is a critical speed map result from the dimensioned drawing of a two-pole induction motor. What can we say about the work of the motor design engineer?

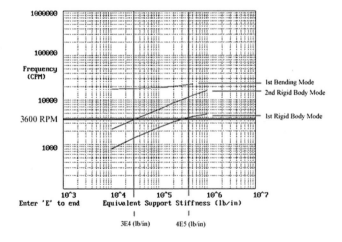

9. What will the following equation yield for the preceding critical speed map?

$$? = \text{motor bracket } 1/k + \text{motor foot support } 1/k$$
$$+ \text{motor foundation } 1/k + \text{subsoil } 1/k$$

10. True or False: A rotating machine with mounting feet that run the entire length of the frame only needs to be shimmed at the four corners.

11. The plot in question 5 has a peak at 7,200 cpm. It appears to almost disappear when the motor power (60 Hz) is cut for the capture of the coastdown. Why?

12. The plot below represents a 2D rotor model for a two-pole motor. What mode is it?

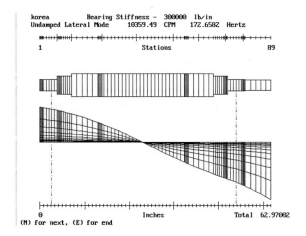

13. Where does this second rigid-body mode appear due to an equivalent support stiffness of 300,000 lb/inch?

14. What is the typical name given to the plot below?

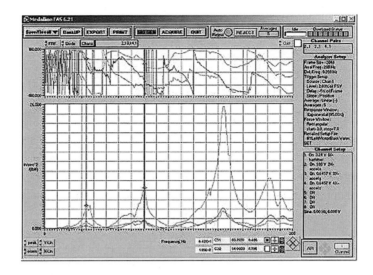

15. What is the purpose of the plot below?

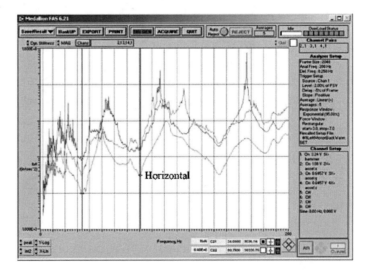

16. Why would anyone design a tuned absorber for a machine?

17. How does a tuned absorber work?

18. A customer installed 18 motors, all driven by VFDs. The manufacturer had given the following engineering instruction for installation: "One set of power leads through any one motor conduit." Below is a picture of the resulting installation. Will there be a problem?

19. When installing a VFD on a motor, what specification, if overlooked, can cause turn-to-turn shorts?

20. The plot below is a FRF taken from impacting and measuring the same point on a structure. It is also known as a *driving-point FRF*. What can we expect to see in the resulting modal analysis animated shapes at this point?

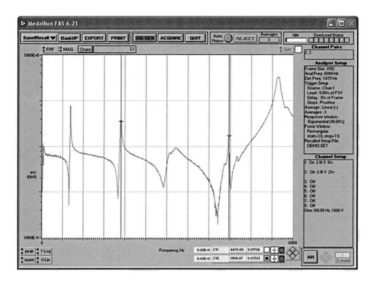

21. What is the most important feature of a driving point FRF?

CHAPTER 8

Electric Discharge Machining Damage

This chapter, after reviews from several prominent people in the industry, has turned out to be the most well-received. I owe a debt of gratitude to Mr. Patrick Link for his original white paper on EDM (see the Note later in this chapter). All I did was follow Mr. Link's advice, to the letter. This resulted in the correction of every EDM problem encountered over a 20-year period.

There has been a sharp increase in bearing damage in electric motors due to electric discharge machining (EDM). Reports of damage in only two of eight identical machine trains is common. Multiple replacements of bearings in the same motor can occur in a matter of months. The end user has a legitimate complaint for not having answers from the OEM community. This has led to a wide variety of remedies that unfortunately are not permanent. This chapter is a compilation of years of technical articles, field-test results, and the formation of basic rules that will yield permanent eradication of damage due to EDM.

Note: I owe a debt of gratitude to Patrick Link for his original white paper on EDM titled, "Minimizing Electrical Bearing Currents in Adjustable Speed Drive Systems" [Institute of Electrical and Electronics Engineers (IEEE) Industry Applications Society (IAS) Pulp and Paper Conference, Portland, ME, June 1998]. This paper not only defined the causes of EDM, but also laid out the testing plan to define and isolate EDM. Patrick Link's techniques and solutions yield complete eradication of the problem.

Before the advent of the variable-frequency drive (VFD), motors operating on 60-Hz AC voltage or rectified DC voltage could experience this problem. The motor is usually larger than a 444 frame. The first two digits in the frame size divided by 4 represents the distance from the bottom of the motor foot to the shaft centerline in inches. Therefore, a 444-frame shaft centerline is 11 inches above the bottom of the motor feet. This frame size is the smallest that can have an EDM problem because of the length of the stator core being susceptible to a voltage imbalance from one end to the other. Any time there is a voltage imbalance, a current path will be created. The problem could happen in both AC and DC motors. Because the frequency of the voltage is low, the solution is to interrupt the low-frequency current path. This path travels from the motor winding, through one bearing to the motor shaft, down the shaft to the other motor bearing, and then back into the motor winding. Insulating the bearing on one end of the motor (preferably the non-drive end [NDE]) will break this current path and eliminate the problem.

The insulation process must be done correctly. Many large motor manufacturers have written specifications for this task. The best approach for insulating the bearing from the motor frame is to bore the frame inside diameter (ID) larger than the bearing, press a phenolic ring into the frame, and then bore the ring back to the original bearing outside diameter (OD) specification (Figure 8.1). This allows for the replacement bearings to be standard. There are ceramic-coated bearings that will work, but they add substantially to the bearing replacement cost.

Figure 8.1　Phenolic ring pressed into NDE bearing bracket.

There is also a need to insulate the sides of the outer race so there is no contact with the inner or outer bearing caps. A good product for this purpose is Garolite G-10/FR4, an epoxy-grade industrial laminate and phenolic. Any ancilliary item that is added to the bearing or shaft, say, a bearing resistance temperature detector (RTD) or a tachometer on a DC motor, also must be insulated, otherwise the path will be reestablished. Variable-speed drives have added an additional level of complexity to the EDM problem.

Reliance Electric modified a motor to monitor shaft voltages and bearing currents in order to capture the current arc that causes EMD (Figure 8.2). The time waveforms in Figure 8.3 show that if the voltage can dissipate through the lubrication to the outer race before reaching a given level, there is no current arc.

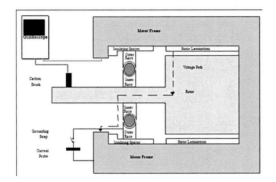

Figure 8.2   Diagram of test motor set up to catch current arc.
(Courtesy of Reliance Electric.)

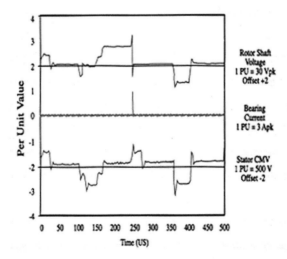

Figure 8.3   Time waveform captures arc in bearing at 250 µs (middle trace).
(Courtesy of Reliance Electric.)

Bearing currents should be less than 0.4 A/mm² (Figure 8.4). This is not the same as the shaft current readings discussed in this chapter. Certain machines can have higher shaft currents without resulting damage due to multiple surfaces to dissipate the current—for example, a motor driving a gearbox through a conducting coupling. The shaft currents will pass through into the gearbox going to ground over multiple contact points. This is not recommended. All stray electric currents have the capacity to cause damage. Also, if allowed to pass onto the ground grid of a plant, they can cause issues with computers, controls, and so on attached to the same ground grid.

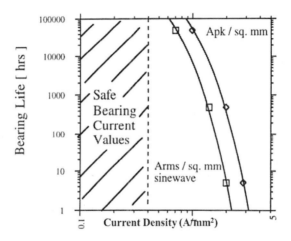

Figure 8.4   Current levels for safe bearing life.

The electric current that causes the damage is due to a voltage potential buildup that occurs between the rotor and the bearing races. There is a capacitance formed by the lubrication in the bearing. When the voltage potential builds to a level that is significant (>30 V), an arc of current will strike the outer race of the bearing and rip metal from it. Note in the time waveform that the rotor had charged to 30 V but did not dissipate any voltage before the next spike from the drive. Typical current arcs of approximately 3 A will tear metal from a bearing race.

CASE STUDY

A customer had its own power plant using four diesel generators. A shutdown was scheduled for normal maintenance. The two running diesels were shut down. The diesel noise disappeared immediately, and all that was left was the coast-down of the generator. The noise from the generator bearing was so severe that all personnel ran from the building. A vibration analysis was scheduled, with the results shown in Figure 8.8.

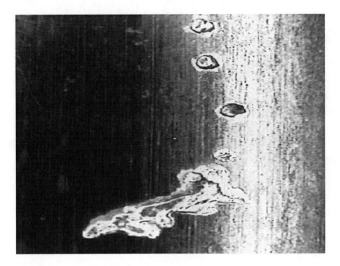

Figure 8.5   Early damage to a bearing outer race from current arcing.

Peak-to-peak acceleration of more than 9*g* was captured (Figure 8.6).

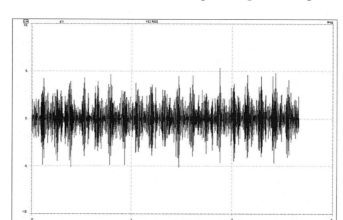

Figure 8.6   Acceleration-time waveform of generator NDE bearing,

Vibration energy indicated that the bearing had failed. The harmonic markers in the spectrum are lined up on a fundamental of 3.08561, the ball pass frequency outer race (BPFO) for this bearing (Figure 8.7). This example also illustrates the frustration some analysts may face in identifying ball bearing frequencies because some geometries will place the BPFO very near harmonics of running speed. The modulation of this frequency helps identify it as the BPFO. The bearing was removed and sent to the motor repair shop for inspection.

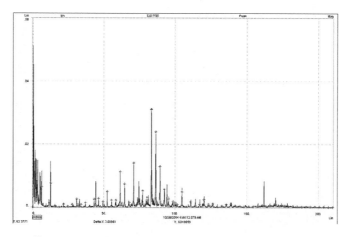

Figure 8.7   Velocity spectrum on generator NDE bearing.

The outer race of the bearing had numerous grooves that had been "machined" by arcing. The even spacing occurs from operating at one speed. EDM related to 60 Hz can be remedied by isolating the NDE bearing from the bracket. The bracket was removed and brought to the motor shop for modification. An inspection of the bracket found a piece of 0.010-inch thin plastic that had been wrapped around the bearing for insulation. Pressing the bearing into the bracket had peeled the plastic strip back, reestablishing the ground path. Unfortunately, the bracket had insufficient material at the hub to allow for machining (Figure 8.9).

Figure 8.8   60-Hz EDM damage.

A new bracket was manufactured by the OEM and shipped to the motor shop. The hub had been sized to permit the insertion of a ¼-inch-thick phenolic ring to isolate the bearing.

Three additional brackets were manufactured and shipped to the motor shop for retrofitting the remaining generator brackets.

The VFD takes 60-Hz AC power and rectifies it to DC on a 600-V buss. The DC voltage is switched on and off at varying times by a bank of high-power transistors (gate turn-off thyristors [GTOs]) at extremely high rates, also known as *pulse width modulation*. The switching occurs anywhere between 12,000 and 20,000 times per cycle! This means that the DC is quickly switched on and off to create the top of the sine wave cycle, and then the switching is slowed, passing through zero, and the polarity of the switching is reversed and then quickly switched again to create the bottom of the sine wave cycle (Figure 8.10). The common-mode voltage is zero for

AC voltage. The common-mode voltage from a VFD output is always ±600 V (Figure 8.11).

Figure 8.9   Original generator bearing bracket.

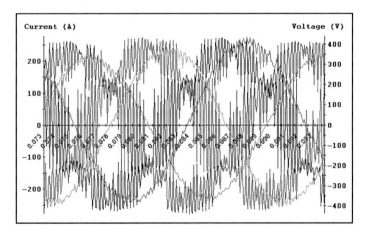

Figure 8.10   Switched DC simulated sine wave identifies switching.

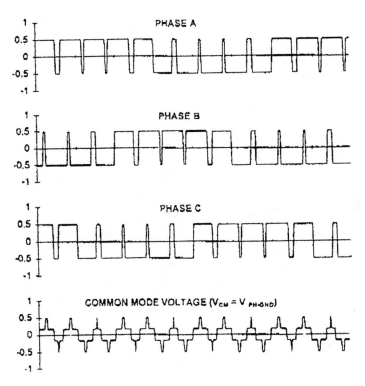

Figure 8.11 Common Mode Voltage on VFD.

CASE STUDY

A manufacturer had four fans on VFDs installed at its facility. The plant was experiencing motor bearing failures, and an inspection discovered EDM damage (Figure 8.12).

Figure 8.12   Installation of fans on VFDs.

EDM damage on VFD-driven motors will have a smearing of the grooves due to arcing occurring at different speeds (Figure 8.13).

Vibration data were recorded on all four fans. The data indicated that one of the four fans already had EDM damage. A test technician was also on site from the drive manufacturer. The test procedure for the drive's technicians included taking voltage readings at various grounding locations to isolate the return paths. Importantly, the drive technicians were not trained in taking current readings.

The plot in Figure 8.14 was compiled from vibration readings on all four motors. Both bearings of each motor are plotted. Note that the number 3 motor exhibits heavy impacting in the time domain and BPFO modulation in the spectra, but the other three motors exhibit no evidence of EDM damage.

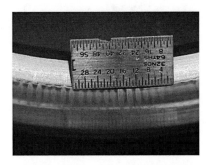

Figure 8.13   VFD EDM damage is smeared because of speed changes.

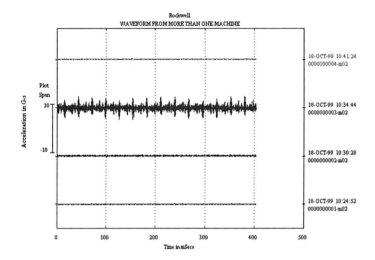

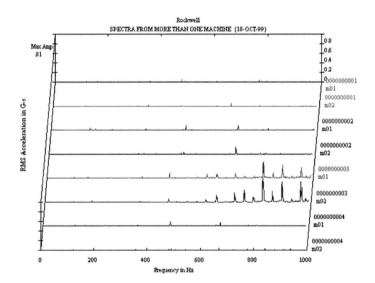

Figure 8.14   Vibration data from all four motors.

The grooves in the outer race cause multiple impacting events as the balls pass over that area. This generates modulation at the BPFO. The result is sidebands at the BPFO frequency that usually appear between 60 and 120 orders of running speed (Figure 8.15).

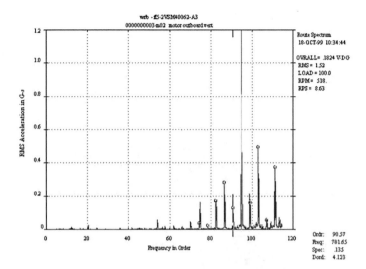

Figure 8.15   Number 2 fan EDM damage vibration.

VFD applications are much more difficult to analyze than 60-Hz low-frequency-current problems. The solution to this VFD problem was not that difficult. An inspection of the VFD control room was made after taking readings on the motors. The power cable was routed from the motors to the room through underground conduits. The power cable came up through the concrete floor and then through the bottom of the VFD cabinets. The installer of the equipment had cut the grounds off below the cabinet floor! There was no connection to any ground in the cabinets.

I had not yet read Patrick Link's paper. Voltage readings were taken that indicated that the shaft voltages were not in tolerance. The high-frequency switching can cause currents ranging from 4 to 10 MHz. These are radiofrequencies and therefore very difficult to measure accurately and sometimes impossible to trace. The proper tool for measuring these high-frequency currents is a *Rogowski coil*.

Power Electronic Measurements of Nottingham, England, manufactures these devices. The unit consists of an air-cored current sensor that has no magnetic materials to limit its ability to identify high frequency currents. The coil conveniently opens so it can be placed around a shaft or power lead and then closed. There is a signal-conditioning box that converts the output signal to a voltage source, with an output sensitivity of 50 mV = 1 A.

ROGOWSKI COIL MEASUREMENTS

A 20-MHz single-channel oscilloscope can be used to read the output of a Rogowski coil. A storage scope should be used to save the data for the final report (Figure 8.16).

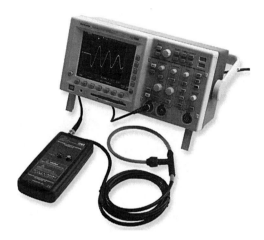

Figure 8.16   Power Electronic Measurements (PEM) Rogowski coil.
(Courtesy of Power Electronic Measurements.)

The key to preventing all EDM issues is to remember that the current created by the VFD must return to the VFD from the motor through the *path of least resistance*. That path can be supplied by the proper installation of the entire system (VFD, power cables, and motor), or it will find its own path. The path chosen by the system for the current return to the drive, when left to its own devices, is *not* a reliable path.

Troubleshooting an EDM problem with a Rogowski coil requires multiple measurements in order to quantify how much common-mode current there is from the drive source and then to determine the paths the current is taking from the motor back to the drive (Figures 8.17 and 8.18).

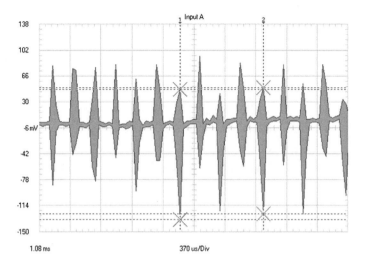

Figure 8.17 Rogowski coil output from a problem motor: shaft current in excess of 3.5 A.

Figure 8.18 Example of a Rogowski coil measuring shaft currents
on a VFD-driven equipment shaft.

These are the measurement point locations for a Rogowski coil (Figure 8.19):

1. **All three phases inside the inverter cabinet.** Measures total common-mode current. Some drives produce more common-mode current than others.
2. **Total conduit and ground just outside the cabinet.** This reading should be zero. Any current measured is returning by unwanted paths.
3. **Total conduit and grounds just outside motor terminal box.** This reading should be zero. Any current measured is returning by unwanted paths.
4. **All three phases and grounds inside the motor terminal box** (not critical).
5. **Shaft of driven unit on units with conducting couplings (non-insulated).** Identifies any shaft currents from the motor that may damage the driven equipment.
6. **Auxiliary grounds.** These can provide a path to ground, but it will be the plant ground grid. It is not advisable to put high-frequency currents on the same ground used by plant digital controls.
7. **Shaft and grounding brush of the motor.** I have found little difference in gathering true shaft current readings with and without the grounding brush. Also, the grounding brush is an added safety issue.

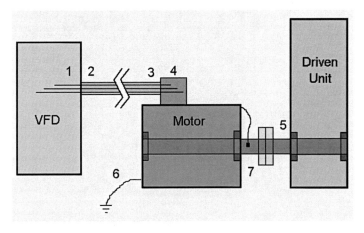

Figure 8.19   Rogowski coil measurement locations.

So how does an end user provide the path of least resistance? Remember that these frequencies are radio frequencies. High-frequency current will not travel well through round solid-copper wire. Therefore, grounding for a motor to a VFD that consists of twisted strands of round copper wire will fail! Look at a commercial high frequency meter. There are no solid wires for the input signal. The input is a square, hollow-copper tube. The high frequencies ride the walls of the tube! How do we duplicate this for our VFD-motor connection? VFD power cables. The entire power lead bundle is surrounded by a copper sheath. The copper sheath is peeled from the power leads inside the VFD cabinet and attached to the protective earth (PE) ground connection. The other end is routed inside the motor terminal box, and then the copper sheath is peeled back, lugged, and bolted to the frame of the motor housing inside the motor terminal box. This copper sleeve will carry the high-frequency current back to the drive.

CASE STUDY

A plastics manufacturer was experiencing multiple motor bearing failures on its main extruder. The VFD power cable was installed after the failure of eight sets of bearings over a one-year period. The OEM had claimed that it had installed shielded cable. The braided shields were separate over each lead, not over all three power leads together (Figure 8.20). This will not work. Shaft currents were in excess of 6 A on this motor. The cable was replaced with a copper sheath–shielded power cable (Figure 8.21).

Figure 8.20   Installation of improperly shielded power cable.

The motor still had one more bearing failure after the cable installation, that was due to low frequency circulating currents. The insulation of the motor NDE bearing was not installed properly.

Figure 8.21   Copper sheath braided and connected to motor frame.

Figure 8.22   Stranded copper grounds and copper sheath joined at the drive PE ground.

So the *11-inch rule*, which is still necessary for motors on VFDs, takes care of the low-frequency current path. The correct power cable takes care of the high-frequency current path. The insulation of the NDE bearing is very important. If this area of concern is suspect, perform the following test:

1. Remove the DE motor bearing bracket while supporting the shaft.
2. Leave the rest of the motor assembled.
3. Support the motor DE shaft, and center the rotor in the air gap.
4. Megger test the motor shaft to the motor frame.

The problems I have encountered over the years have continually reinforced the diagnostic procedures and fixes described by Patrick Link. For example, a large automobile manufacturer was experiencing bearing failures on large press motors. I asked if VFD-duty power cable had been used for the installation. The answer was affirmative. I scheduled a trip to the site, and I spotted the problem immediately.

The coaxial shield had been properly attached to the motor terminal box with the application of packing glands (Figure 8.23). I checked the VFD cabinet next. The braiding had been attached to the 60-Hz grounding Unistrut beam in the bottom of the VFD cabinet but not extended to the PE ground of the drive.

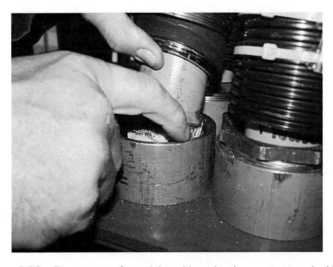

Figure 8.23 Proper use of coaxial packing gland on motor terminal box.

I procured some braided grounding strap from an auto parts store and routed a section of it from the Unistrut grounding beam to the PE ground connection. The shaft currents dropped from 2 A to 0.150 A (Figure 8.24).

Figure 8.24   Addition of automotive braided grounding strap to alleviate shaft currents.

Motor manufacturers get the complaints, yet the drive is the main culprit because of the higher common-mode current. Do not rely on grounding rings, ceramic bearings, and so on to solve these problems. None of these solutions, in my opinion, are a permanent fix.

CASE STUDY

New construction at a large manufacturing facility included four fans. All four fans were VFD applications. The customer complained that the two Reliance motors had high vibration and high noise, whereas the two Siemens motors were quiet.

Vibration readings were taken on all four fans, and the two Reliance motors had bad EDM damage. A Rogowski coil also was used to identify the cause of the EDM. The Reliance motors had more than 6 A of shaft current. Readings on the three power leads inside the VFD cabinet identified the problem. The common-mode current measured on the VFDs was drastically different. The two drives on the Reliance motors had 250% more common-mode current. The two drives on the Reliance motors were "new and improved!" This means cheaper with more common mode current noise.

The project engineer was called for a meeting. The findings were conveyed to him, and his response was a request for Reliance to replace the motor bearings in both motors and then to add a grounding brush to the NDE of the motor. The motors were connected to the fan shafts using Steelflex couplings. These couplings have a coil spring with a thick enamel paint coating on the spring. However, once the paint begins to chip, they become conducting couplings. The project engineer was warned that not replacing the power cable with VFD-duty cable would result in the high shaft currents eventually passing through the coupling into the fan shafts. The project engineer was unmoved. The job was finished as specified, and time passed . . .

The contractor on site called me 2 years later and said, "If you can send me a copy of your final report stating that the shaft brush would not work, I have a picture for you!" A copy of the report was sent, and Figure 8.25 shows the picture I received back from the contractor.

Figure 8.25   Fan shaft portion that rides inside the fan bearing.

The shaft currents were confined to the motor bearings until the paint chipped on the Steelflex spring, and then the 6 A found a ground path through the fan bearings. There is no way to know how long this bearing had experienced EDM.

CASE STUDY

A county wastewater pumping station was experiencing EDM damage to the motor bearings. The customer mounted grounding brushes on one motor and followed my recommendation to replace the power cable with VFD-duty cable on a second motor. The plant manager stayed with the experiment for two years. The motor with the grounding brushes had been through two motor bearing changes. The motor that had the power cable replaced was operating on the same bearings.

Figure 8.26 shows a common mistake when VFD shielded cables are attached (grounded). The braided shield is cut short and not connected to the VFD PE ground terminal. A high-frequency current measurement of the unit indicated a 5-A differential between the clamp point and the PE ground terminal. The impedance of all the "bolted" connections to get to the PE ground is the issue.

Figure 8.26   Braided shield bared and clamped to the mounting plate.

The shaft currents were measured at 1.48 A without properly connecting the shields (Figure 8.27).

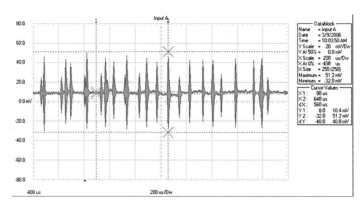

Figure 8.27 Shaft currents with improperly grounded shields.

The shaft currents dropped to 0.48 A when the braided shields were extended from the clamp connection point to the PE ground terminal (Figure 8.28). A more thorough job of running the braided shields was completed a month later with the result that shaft currents dropped to 136 mA. I have estimated through numerous measurements in the field that any shaft current readings less than 0.5 A will not cause bearing damage. Readings as high as 1.0 A have been known not to cause damage but are not recommended.

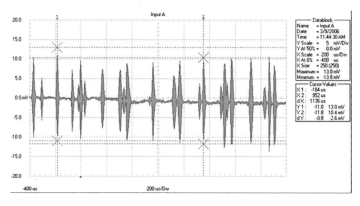

Figure 8.28 Extended shielded braiding attached to the PE ground terminal.

CASE STUDY

A customer was experiencing multiple motor bearing failures on its chilled-water pumps. Testing with a Rogowski coil identified high-frequency current on the motor shaft. The recommendation was made to replace the VFD-to-motor power cable with VFD-duty cable. The cable was installed, and retesting was scheduled. The testing identified low-frequency currents on the motor shaft, indicating that the NDE motor bearing was not insulated. The customer acknowledged that its previous motor repair shop had insulated the NDE bearing. The motor bracket was removed, and the bearing was wrapped with a 0.010-inch piece of plastic. Also, there were no insulation disks between the bearing outer race and the inner and outer bearing caps. The motor end bracket was shipped to my motor shop, the bearing bore was opened an additional ½ inch, and a phenolic ring was inserted. Two disks were machined to isolate the bearing end faces. The motor was reinstalled, and shaft current readings were now in the 100-mA range. Their motors have not been removed for EDM damage in more than six years.

CASE STUDY

A paper mill had large VFD-driven fans. The OEM had installed grounding rings on the shaft next to the motor DE bracket. The shaft beneath the brush had turned blue, and sparks were visibly jumping from the motor bearing bore to the shaft. The shop sent a technician to measure the shaft voltage with a warning to be aware of high voltage levels. The shaft voltage measured 490 V. The motor was taken out of service, and during disassembly, it was discovered that the OEM had used too long a bolt for the end bracket during manufacturing and had driven it into the end winding. The customer was also advised to check the VFD. The VFD should have tripped with 490 V going to ground.

Note: I have relied on extensive testing by various motor manufacturers to quantify the ability of brush rings to permanently eliminate EDM issues. No motor manufacturer, as of the date of this writing, is willing to place any grounding brush or brush ring on its product and warranty it against bearing failure due to EDM.

REVIEW QUESTIONS

1. True or False: Only electric motors on variable-frequency drives (VFDs) have electric discharge machining (EDM) damage.

2. True or False: The damage shown in the image below, due to EDM, takes several years to occur.

3. The velocity spectrum below is due to EDM damage. There are multiple sidebands in the spectrum that are usually separated by what frequency?

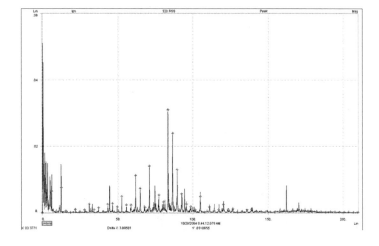

4. The plot below is voltage and current from the output of a VFD. What can you see in the voltage traces that defines them as a VFD source?

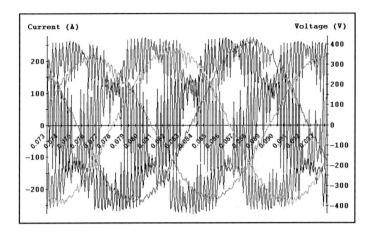

5. The following plot of the voltage output from a VFD illustrates why the total 600 V is always present in the common-mode voltage from the source. What is the common-mode voltage of utility power?

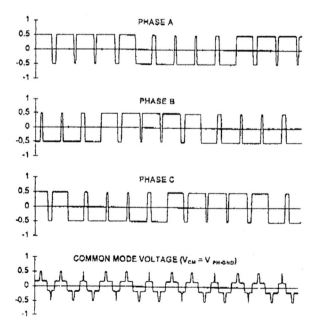

6. The following photograph of EDM damage on the outer race of a bearing appears to be smeared. What could be the cause of this pattern?

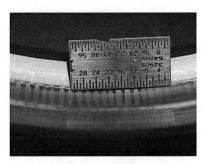

7. The following stacked time traces from four separate motors identify one as experiencing impacts in the ball bearing. What is the separation in the impacts in this time waveform?

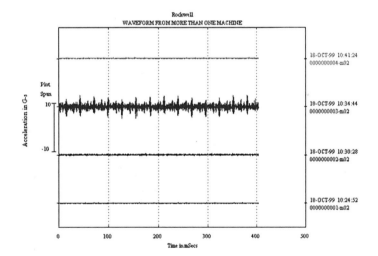

8. Name the measurement tool seen in the following photograph.

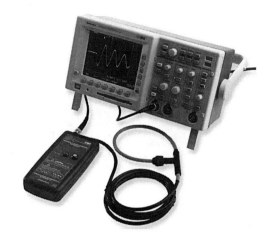

9. Readings taken with this particular model yielded 50 mV/A. Based on this sensitivity, is the motor upon which the following readings were recorded in trouble?

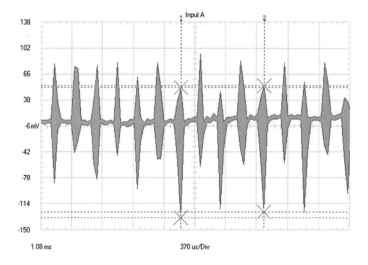

10. An OEM assembled drive assemblies for its pumps. The following photograph was taken on one of the manufactured drive cabinets. The power cable between the motor and the drive was grounded by stripping away the outer cable insulation and then bolting a metal clamp to the backplate of the VFD cabinet. Is this a good installation?

11. According to numerous tests performed by me, what is the maximum amount of shaft current that a motor can operate with and not have EDM occur?

12. True or False: Motor manufacturers will warranty their motors against EDM damage with the addition of shaft grounding brushes.

CHAPTER 9

Electrical and Vibration Testing to Solve Motor Issues

The process of troubleshooting a machine problem needs to include all aspects of the operation. A motor driven by a variable-speed drive adds another dimension that should include using other analysis tools, such as a megger and a micro-ohmmeter, and can require a means of recording and storing current and voltage outputs.

ENCODER ISSUES

CASE STUDY

A sand quarry was experiencing high erratic vibration on its main bucket. The operation included a fast decent of the scoop to the lakebed, where it would take a bite. The ascent included a slow start with a fast acceleration to top speed until the bucket cleared the water. The job was to include a vibration check of the motor. A drives technician had already tested the drive and found no issues. The vibration test of the motor indicated high nonperiodic impacting vibration, especially during the high-speed ascent of the bucket. A static motor circuit test indicated that all phases were well within acceptable specifications. The motor and drive static tests found no problems. What's left?

The technician decided to use a multichannel digital signal processor with the motor circuit tester current probes as the inputs. The motor power requirements were high enough to call for dual power heads. This meant that it was necessary to test six power output leads, drive 1 phases A, B, and C, and drive 2 phases A, B, and C. The probes were placed in various combinations on these six leads to assess the power output waveforms.

The same phase on both drives is faulty *and* identical (Figure 9.1).

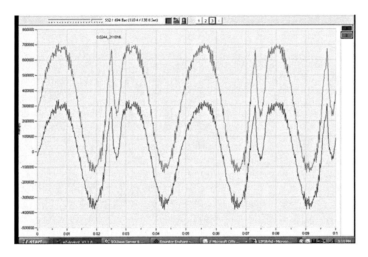

Figure 9.1   Phase A from drives 1 and 2.

The same fault is present simultaneously on phase B on drives 1 and 2 (Figure 9.2).

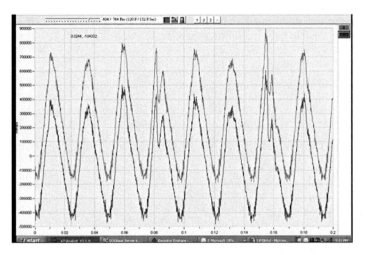

Figure 9.2   Phase B from drives 1 and 2.

Again, the same fault appears simultaneously on phase C (Figure 9.3).

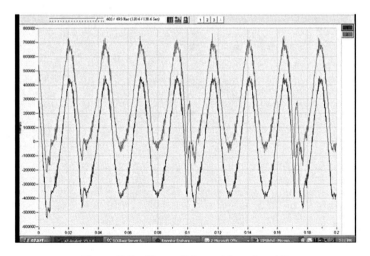

Figure 9.3   Phase C from drives 1 and 2.

The motor vibration on a VFD speed of 42 Hz is 0.15 inches/s (Figure 9.4). This is not excessive.

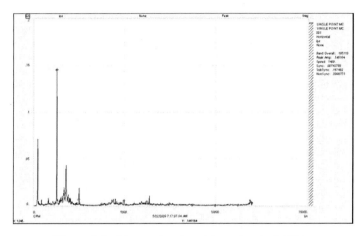

Figure 9.4   Motor vibration at top speed of 42 Hz.

A violent rumble was heard in the motor when the bucket reached high speed on the ascent (Figure 9.5). The base of the predominant peak, at 200 cpm, which is not running speed, has an indication of resonance.

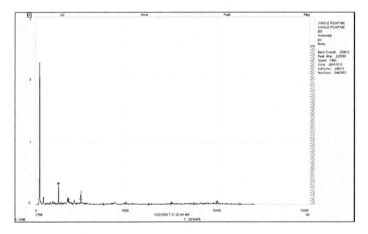

Figure 9.5   Unidentified vibration peak at 200 cpm.

The only components that are common to the motor and the drive that could cause the faulty current waveforms is the tachometer (encoder) attached to the non-drive end (NDE) of the motor shaft and the comparator. The comparator checks the requested speed from the drive against the motor speed returned by the tachometer.

Tachometers are electromechanical devices that are susceptible to failure. The final report called for a check and replacement of the tachometer. The tachometer was replaced, and the problem resolved. The point of this example is the need for a field service engineer or technician to consider all components of the system. Don't assume that testing the motor and the drives separately and finding them acceptable is sufficient.

How much vibration can a bad or faulty signal from a tachometer cause?

CASE STUDY

I was sent to troubleshoot a DC motor vibration problem. The motor in question was the feed motor on a metal shear. Sheet metal was fed along a bed by a large rubber-coated wheel on the motor shaft. The customer said that initial operation of the drive was so violent that all the workers ran from the building. The drive manufacturer had sent a drive technician to investigate, and the technician had spent several weeks adjusting filters on the drive and attempting to dampen out the response, all with limited success. The rumble persisted.

The customer put the machine into operation, and the motor shaft would rotate at approximately 1 Hz for four or five revolutions, and then it would stop, and the shear would cut the sheet. This action would then repeat. How can a motor that turns at 1 Hz vibrate excessively? The customer asked for patience. After five or six sheets had been cut, a rumble was heard from the motor.

I sat down with the drive technician in the customer's cafeteria. The following question was asked several times: "Where does the DC drive get its reference signal from?" The drive technician would always answer, "The tachometer." This is not correct. The tachometer signal to the comparator tells the drive how fast the motor is turning. Because this operation was more concerned with making accurate length measurements, this would require an encoder instead of a tachometer.

The drive technician's eyes lit up, and he exclaimed, "Oh, there is an encoder wheel riding on the steel sheet!"

We both went back out to the line, and there it was. The encoder was mounted on a 6-ft vertical pole. That pole was attached to a 12-ft box beam that spanned the feed bed. At the sides of the bed were 8-ft-high vertical box beams supporting the horizontal span beam. I grabbed the encoder and lightly pulled it horizontally along the bed. The encoder easily moved an inch. Then I released it, and it vibrated like a low-frequency tuning fork for a long time. There was the source of the vibration. The motion of the sheet metal would eventually excite the natural frequency of the encoder support beam. Once the vibration started, the encoder wheel would begin to oscillate. This oscillation was interpreted by the comparator as motor speed fluctuations. The comparator would then send signals to the power head of the DC drive to compensate for these fluctuations. This would cause even more resonant vibration of the encoder support beam. The recommendation was to remove the overhead support system, mount an I-beam to the plant floor at the side of the bed, and attach a horizontal support box beam just above the bed floor that was just long enough to reach the sheet metal. The encoder wheel would then be attached to the end of the beam.

The plant manager came out to hear my explanation of the problem. He replied, "We will try your recommendation. When that doesn't work, you'll hear from me," to which I replied, "And if it works, I'll never hear from you." The manager abruptly replied, "Oh no, if it works, I'll write you a letter." It must be lost at the mail sorting center . . .

I get it. Once one fire is put out, a manager is on to his next fire. It's just what happens.

This is a perfect example of missing the smaller details. The motor and drive are highly sophisticated pieces of hardware. Once that encoder is placed into the control circuit, it is just as important as any other component and requires just as much attention to detail. The encoder housing must be held rigid so that only signals from the motion of the wheel are interpreted as speed changes or angular positions. Remember, those small encoder signals are going to be greatly amplified by the power head of the drive.

VFD ONLINE TESTING

A transportation hub in a major city utilizes supply and exhaust fans to prevent a buildup of bus exhaust fumes inside the building. The fans are 400-hp vertical motors on variable frequency drives (VFDs). The motor windings are electrically tested once a year, and the drives were tested with a PdMA MCEMAX tester every six months.

An EMAX test on one of the drives indicated a high current imbalance between the three phases (Figure 9.6).

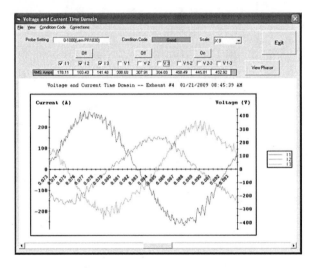

Figure 9.6 Current imbalance.

The three phases of current are as follows:

Phase A: 178 A
Phase B: 103 A
Phase C: 141 A

These are nowhere near acceptable. A 4% imbalance is considered the maximum allowable.

The power results summary page (Figure 9.7) indicates that voltage imbalance is good (EMAX VFD voltage readings will always be erroneous. Percent imbalance is the key and that it is acceptable), but current imbalance is very high. The recommendation was to check the motor leads inside the motor terminal box. Inspection of the leads identified a burned connector. The cause was loosely bolted connections at the lugs. This fault also caused excessive vibration at 2× line frequency at the motor bearings.

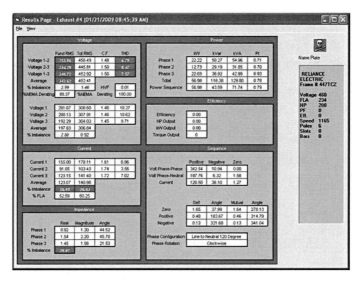

Figure 9.7   Current imbalance of more than 26%.
(Courtesy of PdMA Inc.)

ONLINE TESTING OF LARGE-HORSEPOWER, HIGH-VOLTAGE MOTORS

A power utility has its gas turbine starter motors tested yearly. This includes a complete power study, as well as a current inrush test. The current inrush test can identify problems with an induction rotor or drops in incoming power by capturing the total time it takes the motor to attain nameplate speed. The motors are 2,500 hp, 4 kV.

Caution! Electrical testing of motors has hazards that can kill an untrained technician. Never attempt these tests without a completed work scope of any and all safety items. The online testing equipment is limited to 600 V. Testing online at 4 kV requires installed potential transformers. These transformers are on the order of a 40:1

ratio. This is the only place voltages should be tested with electrical test equipment. Arc flash is another hazard. Usually, plant personnel are trained and equipped with proper personal protection equipment (PPE) to perform the lockout/tagout (LOTO) procedures. If they aren't, again, the technicians require full training in the process and all proper PPE before attempting these procedures.

CASE STUDY

The testing identified a current imbalance on one of the starter motors. This was reported with a recommendation to check all power leads and connections from incoming to the motor. The next time the testing was performed, all results were well within specification. The customer was asked at the completion of the test what maintenance it had performed. The technician had found brittle leads on the incoming power leads from the transformer.

The customer usually will not convey corrective actions taken based on the technician's report. Therefore, it is the job of the technician to gather that information (Figure 9.8 and 9.9).

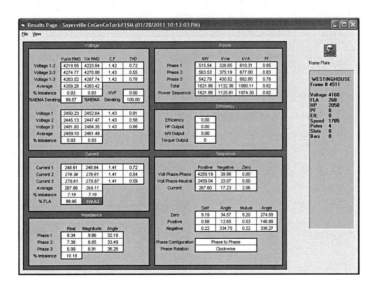

Figure 9.8   Identified current imbalance.

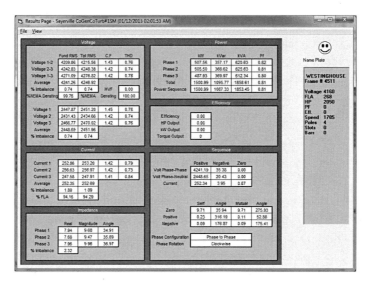

Figure 9.9 Follow-up test after brittle leads were replaced.

COMPLETE ELECTRICAL TESTING TO UNCOVER PROCESS-RELATED MOTOR FAULTS

CASE STUDY

A city hospital was installing an air exhaust system on its roof. A large stainless steel box plenum had two vertical motor-fan units installed on top. The plenum intake was on the end and was connected through a long vertical duct to the floors below. The intakes on the individual floors had not been installed, so the entire side of the plenum had been removed for fan testing. Both motor/fan units exhibited good static test results (Figure 9.10).

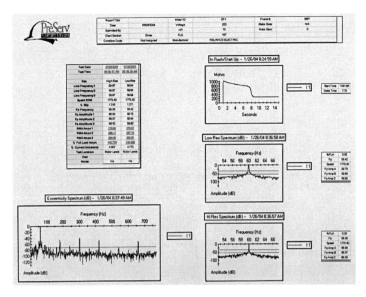

Figure 9.10   Static test results.

The other unit would trip on overcurrent after 20–30 seconds. The contractor wanted to remove the motor and have it shop tested. The logistics to remove this motor were formidable. The plenum was extremely slippery, because it had snowed on the testing day. The motors were in top hats mounted on top of the plenum. There was no construction scaffolding, so all of that would have to be brought in and erected. I asked the contractor to let me finish the testing.

The fans were both tested for motor circuit integrity and found to be within specifications. The Emax® testing provided a complete test for the one motor-fan unit that would stay online. Testing proved the unit ready for full-time operation (Figure 9.11). A current inrush test on the unit that was tripping found an anomaly (Figure 9.12). The unit took more than twice the time to come up to speed. That one item gave the contractor enough information to doubt the way the fans were being tested.

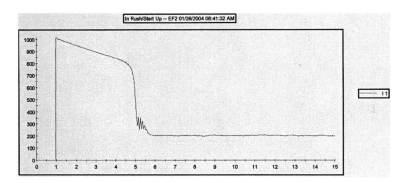

Figure 9.11 Current inrush on good unit.

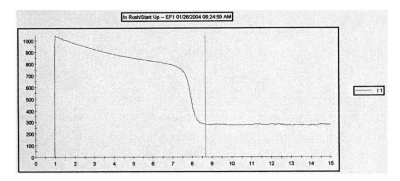

Figure 9.12 Current inrush on faulty unit.

We walked out onto the roof with the good fan in operation. The contractor picked up a handful of snow (how convenient!) and tossed it into the open side of the plenum. The snow immediately went into a circular swirl and went right up the fan intake. He then had an assistant start the other fan. He threw another handful of snow, and it went right past the fan intake and stuck to the back wall! The cutout in the plenum and the air intake to the plenum were creating turbulence and overloading the motor!

The decision was made to wait until later in the week when the plenum side would be closed and the lower-floor intakes were all in operation. The decision proved correct. Both fans worked well.

BROKEN ROTOR BARS

I wish I had a quarter for every time the initial diagnosis was for a motor rotor with broken bars. Knowing how to differentiate between broken bars and other electrical motor vibration is important and can save a technician the embarrassment of having a motor pulled for no reason.

First, consider that there are basically two different types of induction motor rotor designs. There are rotors that have aluminum or copper bars placed inside the lamination slots, and there are rotors where the end rings and bars are cast aluminum. The casting process occurs with the rotor heated and held vertically as the melted aluminum is force-fed up through the core. The copper and aluminum barred rotors have the bars brazed into the end rings.

Reliance Electric has had a design for many years that is referred to as a *loose-bar construction*. Typically, rotor bars are driven through the lamination stack for a very tight fit. The bars loosen up over time after repeated expansions and contractions and are made secure again by swedging the bars. This is a process that uses a small metal wheel in a press. The wheel applies pressure on the part of the bar that is exposed between the edges of the lamination slots. The wheel is then driven down the entire length of the rotor body, causing the rotor bar to blossom in the rotor slot, thereby firming up the fit.

It needs to be said here that rotor engineering designs are solid and very reliable. When a rotor fails, the reason is usually poor construction or misuse.

Do *not* swedge the bars in a Reliance Electric rotor with a loose-bar construction. You will be able to tell a loose-bar rotor by the existence of three copper pins, drilled into the rotor body laminations at the midpoint, every 120 degrees around the circumference. Each copper pin is then brazed to the closest rotor bar. This anchors the entire bar cage axially. Swedging these bars will negate their ability to expand and contract thermally in the axial direction, causing them to bow the rotor. The result will be a large rotor unbalance due to the bowing and the cost of a new rotor.

CASE STUDY

A pipeline station positioned a new motor on a pump and started it up. The NDE motor bearing failed in seconds. Remember the alignment topic concerning limited end float?

The motor was sent to a local motor shop to have the bearings replaced. The repaired motor was put into service, and vibration levels increased over the next hour until the unit was shut down. A motor that ran successfully at the factory would not run smoothly after having the bearings replaced.

The cause of the original failure was the lack of limited end-float buttons inside the coupling. A review of the specifications indicated that the coupling had ½ inch of end float and the motor had 0.450 inch of total end float. The motor shaft was centered within this total, so the shaft shoulder was only 0.225 inch from the bearing. The customer had two other motors in service that had never had this problem. The reason? Those motors had an end float of 0.750 inch. No sleeve-bearing motor should ever be put into service without the coupling end play limited by the provided steel end-float buttons inserted in the coupling. The remaining end float should be on the order of 0.020–0.050 inch.

The repaired motor was started up, and I found very low vibration readings. The customer warned me that they would not remain low. After a one-hour run, the vibration readings were no longer acceptable, approaching 0.25 inch/s velocity.

The motor was sent to the nearest authorized motor repair shop for the OEM. The technicians there ran the motor and claimed it required balancing. I asked that they not do anything until the itemized repair list came in from the local motor shop that the customer had used.

The itemized list finally came, and number 6 on the list was "Swedge rotor bars." Some shops not familiar with this type of rotor construction will have concerns when the rotor is placed in the balance machine. The rotation of the rotor will cause the loose bars to ping, and the conclusion—the *wrong* conclusion—is that the bars are loose, and the shop will swedge the bars.

I called the Reliance Electric shop and asked if the technicians had taken the rotor out of the motor. They had. I asked them to examine the exposed portion of the bars. They could see a deep groove running down the length of every rotor bar.

The rotor was useless. The customer was given a new rotor at a cost of $16,000. The adder for the swedging operation? Four hundred dollars!

Now back to broken rotor bars. Many electric motor shops will use a device called a *growler* to test for broken bars. A growler is an AC transformer device that imparts current into a single rotor bar. A hacksaw blade is convenient to use, and it's placed down along the rotor bar being excited. If the rotor bar carries the AC current, the hacksaw blade will vibrate up and down against the rotor body.

There is a major fault with this kind of test. A rotor bar may, while the rotor is static, still make contact at the break and conduct enough current to excite the blade. However, at speed, the break may open or the connection point may not have enough surface area to carry the necessary current, and that rotor will not operate properly.

The best way to diagnose the condition of an induction rotor is during full-load operation. An induction rotor with a bad or broken bar that is not carrying enough current will fall back on itself rather than continue to move in the direction of the power in the stator. The fault will cause multiple sidebands at the pole pass frequency, which is the slip frequency times the number of poles. For example, a two-pole motor has a field that rotates in the stator at 60 Hz, or 3,600 rpm in the United States. Other countries on 50-Hz power will use that frequency and 3,000 rpm. An induction motor makes torque through the slip or pull of the stator field. A typical slip frequency for a two-pole motor would be 15 rpm. So a two-pole motor would have a running speed of 3,585 rpm. The slip multiplied by the number of poles (two) yields 30 cpm, and that would be the modulation frequency occurring in the rotor due to a broken bar. No matter how many bars may be damaged, the formula remains the same. However, the slip-frequency sidebands will grow in magnitude and number as more bars fail because the rotor will lose torque-carrying capability. This is also a key item to look for if broken bars are suspected. The revolutions per minute of the motor will not match the nameplate specification. This will also increase the slip frequency slightly.

VIBRATION AND CURRENT SIGNATURE ANALYSIS OF A BAD ROTOR

C A S E   S T U D Y

A major HVAC contractor had a chiller that it suspected had a bad rotor. This chiller, 1,750 hp, was 20 ft in the air, and changing the motor out would have been very costly. The data had to be conclusive to make the call.

There were fluctuations in the current meter of the motor control center (MCC). The current fluctuations were approximately 20% at full load. This is a plus for analog as opposed to digital current meters. The damping in the analog meter allows for a clear indication of the amperage swing.

Vibration data and current signatures were recorded at low load (30%), medium load (60%), and the maximum load achievable under the conditions (>80%). The data clearly indicate the need to assess rotor issues at 70% load or greater. The U.S. Navy published a specification in the 1950s that stated that all rotor assessments had to be made with greater than 70% nameplate current applied.

The vibration data indicated no pole pass sideband energy at low load (Figure 9.13). The vibration level is slightly elevated, but there are no other issues within these data.

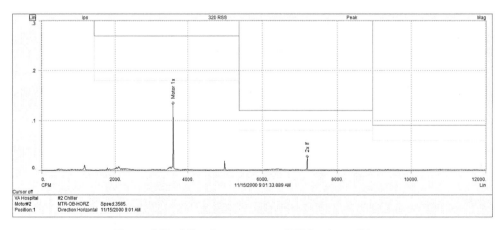

Figure 9.13   Vibration spectrum: 30% load conditions.

The current signature was recorded by taking the analog output of a portable current clamp directly into the handheld data collector (Figure 9.14). The rule of thumb for any sidebands of the 60-Hz peak states that they should be at least 45 dB below the 60-Hz peak magnitude. Twenty decibels is equivalent to a 10:1 ratio. Therefore, the sidebands in Figure 9.14 are more than 60 dB down from the 60-Hz peak. Looks good!

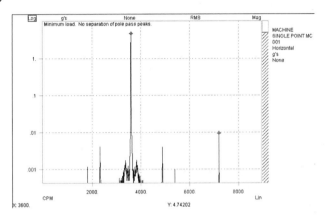

Figure 9.14   Current signature: 30% load conditions.

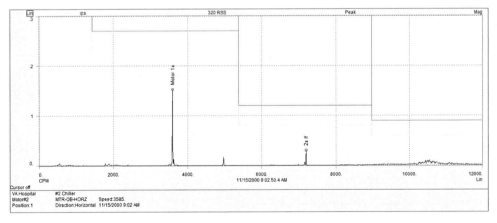

Figure 9.15   Vibration spectrum: 60% load conditions.

At 60% load, the vibration is still slightly elevated. Sidebands are just starting to appear on the sides at running speed 1× and 2× line frequencies (Figure 9.16).

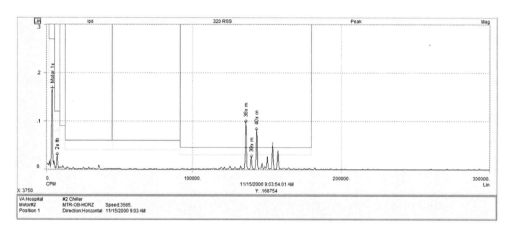

Figure 9.16   High-frequency vibration spectrum: 60% load conditions.

Many vibration manuals call for a quantification of the amount of energy at rotor bar pass and the 2× line frequency sidebands. This energy can be misleading because dynamic rotor deflection in the air gap also can cause this energy. This information alone is *not* indicative of broken bars.

At 80% load, there is now a clear indication of plus and minus pole pass sidebands surrounding 1× and 2× operating speeds (Figure 9.17). These plots should be viewed in a vertical log scale format for easier identification of these sidebands.

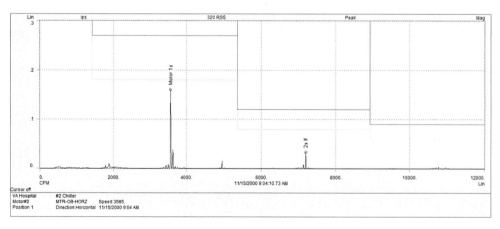

Figure 9.17   Vibration spectrum: 80% load conditions.

A high-resolution plot of vibration in a log vertical scale format clearly identifies plus and minus pole pass frequency sidebands. Slip is approximately 30 cpm; therefore, a two-pole motor would have 60-cpm sidebands. Note that the sideband cursor readout is "Delta X: 67.5'."

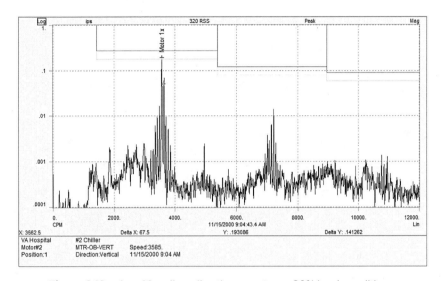

Figure 9.18   Log *Y* scaling vibration spectrum: 80% load conditions.

At 80% load, a major issue with this rotor becomes apparent. The sidebands are only 24 dB down from the 60-Hz peak (Figure 9.19). This rotor, based on the previously gathered data, was called out for replacement.

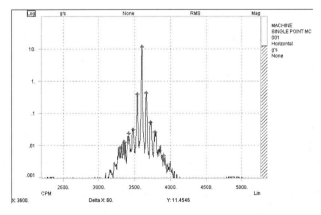

Figure 9.19   Current signature: 80% load conditions.

DESTRUCTIVE TESTING OF A ROTOR

Rotor inspection indicated high heat in the area of the bars near the end rings. The aluminum has melted and pooled on top of the laminations (Figure 9.20).

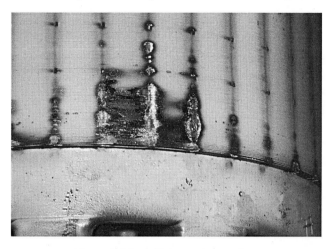

Figure 9.20   Initial rotor inspection.

The rotor was destructively inspected in a lathe to find the root cause. Porosity exists at the junction of the bar-to-end-ring location. Over time, this area had sections of the bar carrying the current through too small of an aluminum cross section, causing the aluminum to get hot and break down. Eventually, this rotor would not have been able to generate the required torque (Figures 9.21 and 9.22).

Figure 9.21   Internal view of the ends of the cast rotor bars.

Figure 9.22 Close-up of voids in the aluminum bars.

KEY POINTS ON TESTING TO IDENTIFY A BAD ROTOR

- Use high resolution for vibration and current signatures, preferably 3,200 lines over a 12,000-cpm frequency range.
- Be sure that the motor load is at least 70% of the nameplate rating.
- Clearly identify the slip frequency.
- Verify that all parameters point to a rotor problem *before* pulling the motor:
  - Does the current signature indicate a rotor problem? (First slip sideband to the left of 60 Hz less than 40 dB down from 60-Hz magnitude.)
  - Does the vibration signature indicate a rotor problem? (Pole pass sidebands surrounding 1×, 2×, 3× speeds and 2x line frequency.)
  - Does the *analog* current meter on the MCC fluctuate? (Greater than 5% of total reading.)

(This can also be done by passing 70% nameplate current through a single phase of a stationary motor and watching fluctuations on an analog current meter as the rotor is turned.)

Also, there are other factors that alone may not point to the rotor but can accompany the preceding conditions, such as motor revolutions per minute not meeting nameplate in order to meet load demands and excessive heating of the motor.

The cause of the failed rotor in Figure 9.21 was the formation of gas pockets at the connection point between the bars and the end ring. Placing vents in the laminations allows the gas to escape, preventing the voids (Figure 9.23).

Figure 9.23   Vents in the laminations prevents pockets in end rings.

How can the integrity of a cast rotor be confirmed if there are no loaded vibration or current data? A current induction test coupled with infrared thermography works well.

The thermographic image in Figure 9.24 clearly identifies an area near the rotor bar–end ring connection point that is conducting more heat than the rest of the rotor or end ring. This area has a high probability of having porosity.

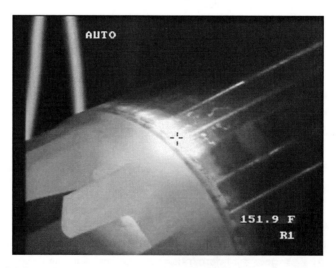

Figure 9.24   A thermographic image of induction testing of an induction rotor.

I was recently involved with the verification of a bad aluminum cast rotor by numerous testing methods. The full load vibration data from the field identified a bad rotor by exhibiting multiple pole pass sidebands. The motor was brought to the Integrated Power Services shop in Rock Hill, South Carolina. The motor was tested utilizing multiple methods, including single-phase testing at 80% load, dyno load testing, and a commercial core-loss tester. None of these tests provided a conclusive answer. I was convinced from the field load data that the rotor was flawed.

Charles Blankenship, an engineer with IPS, finally provided proof. The core-loss tester was utilized to induce a high amount of current directly into the aluminum end ring.

Magnetic Paper was placed on the energized rotor and the "missing" aluminum was identified!

Figure 9.25   Core loss tester clamps attached to aluminum end ring.

Figure 9.26   Missing aluminum bar identified by Magnetic Paper.

CASE STUDY

A database center requested a vibration analysis on a chiller motor that was suspected of having a bad rotor. This rotor was in such bad condition that it took several seconds after the start button was pushed for the rotor to start turning. The unit just sat there and growled. It finally came up to speed, and the vibration readings indicated the most severe case of modulation due to broken bars that the technician had ever witnessed (Figure 9.27). This was the second bad rotor at the same facility. A discussion with the HVAC contractor uncovered the problem. The building temperatures were adjusted by turning chiller units on and off all day long. One of these chillers was being subjected to 10–15 starts an hour. The heat that is generated in an induction motor, above National Electrical Manufacturers Association (NEMA) frame size, is substantial, and for this reason, starts should be limited to two per hour on a cold motor and one per hour on a motor already at operating temperature. Most large motor controllers have dedicated controls to limit starts to these specifications. If the controls have been modified, the motors are going to fail.

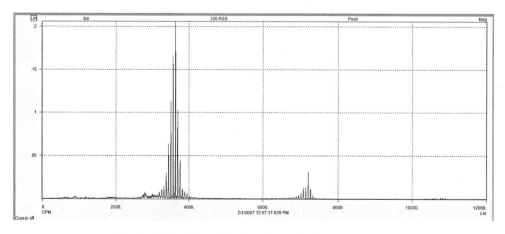

Figure 9.27 Motor vibration due to broken rotor bars.

The solution to this problem of better temperature control is the addition of a smaller screw compressor chiller. These units can be cycled on and off all day long with no loss of integrity.

REVIEW QUESTIONS

1. What can be inferred by the following voltage trace from the same phase A but from two separate power heads?

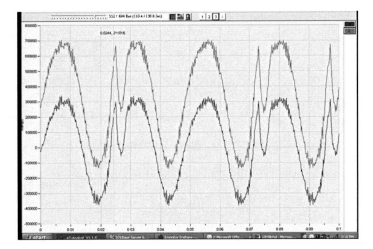

2. Very high vibration levels were experienced on a drag line when the bucket, full of wet sand, was shifted into high speed for the ascent from the lake. The frequency of this peak is much lower than the running speed of the motor. The resulting vibration signature was captured in the following image. What could cause this low frequency peak?

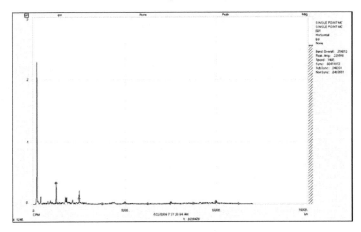

3. A DC motor-drive system has a comparator. The comparator's job is to compare the existing speed of the motor with the requested speed. The motor speed is transmitted from a tachometer on the outboard shaft of the motor. The requested speed can be from a programmed source, or it can be an analog signal from the application itself. What is the typical component that supplies this signal?

4. Encoders can have very high tachometer pulse rates, on the order of 1,024 pulses per revolution. What is critical about the placement of an encoder?

5. The following PdMA EMAX® plot was recorded on the output phases of a variable-frequency drive (VFD). What can the reader infer from this plot?

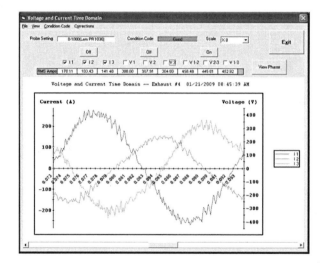

6. The industry standard for balanced voltages in three phase systems is 2%. When the output data is taken on a VFD, the voltages are not easily read. What should be the current imbalance based on a specification of a voltage imbalance of 2%?

7. This is the results page of the PDMA EMAX® test of a VFD. Note that the voltage readings will always be erroneous because of the nature of the "chopped" DC that forms the voltage output. But what can the reader say about the current readings?

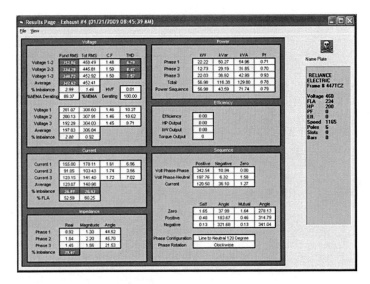

8. Below are the power results of a large, 2,500-hp motor. What is the problem?

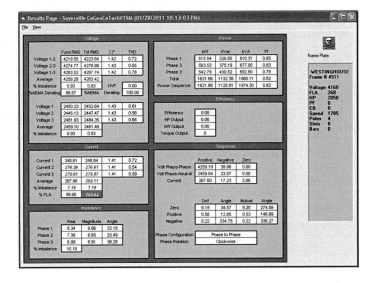

9. Two identical fans were tested for motor inrush current. One motor would trip after taking too long to achieve full speed, whereas the other motor exhibited no issues. See below:

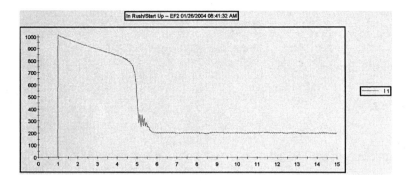

Current inrush on good unit.

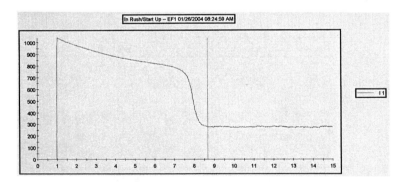

Current inrush on faulty unit.

Other than the time required to achieve full speed, is there anything else the reader can confirm from this test?

10. Broken rotor bars are not a common fault. What are two main causes of broken rotor bars?

11. True or False: Broken rotor bars are just as easily identified at rest as during full-load operation of the motor.

12. True or False: Faulty poured aluminum bars are easily found during static testing.

13. What can you do with a Reliance Electric loose rotor bar construction on which someone has swedged the bars?

14. What are the three items that should be verified before condemning a rotor in the field as having bad rotor bars?

15. I was sent to South Korea to investigate a problem with a two-pole motor. The vibration signatures indicated an increase in 1× vibration as the motor, under load, came up to operating temperature. Unfortunately, a faulty rotor was the diagnosis. This was a cast rotor. What two tests could have been performed that would have diagnosed this problem *before* shipment?

CHAPTER 10

System Design for Measuring Vibration Prior to Shipment

A finished machine's operating conditions, whether from an OEM or a repair shop, should be documented and reported to the customer. This information can be extremely valuable if the end user is not content with the "final product" as it operates in the field. This data can sometimes be the only evidence that a poor running machine is not the responsibility of the OEM or repair shop.

I designed a test stand vibration system, hardware, and software utilizing National Instruments products. The main goal was a system that was fast, easy to use, and yet powerful enough for an engineer to view the post test data if any issues were present. This system, as designed, gives the test stand technician a live view of the velocity spectra of both ends of the machine in three axes: horizontal, vertical, and axial (Figure 10.1). The lower left window is a display of the acceleration-time waveforms of all six measuments. This view is helpful in identifying impacting events occurring in the bearings of the machine. A newly refurbished ball bearing motor will have levels in the 0.4g–1.0g peak-to-peak range.

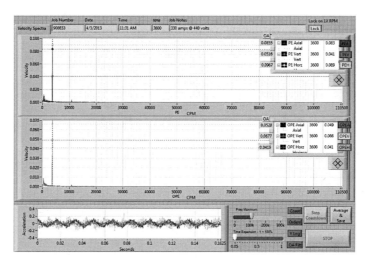

Figure 10.1   LabVIEW software designed to measure machine vibration.

Do *not* test motors with cylindrical roller bearings on the test stand. Roller bearings, by their very design, have excess clearance over ball bearings. This clearance issue is not a problem once the radial load is applied to the bearing. For example, roller bearings are used extensively on motors for belted applications. If there is no belt load, the rollers will skid. Any run time on these bearings without proper load is detrimental to bearing life and should be avoided.

The frequency and time axes of all three windows are adjustable. There is a display of the overall velocity levels for each channel. If the overall level is much higher than any of the peaks in the spectrum, this is an indication that the frequency range should be expanded to find where that energy resides.

A button allows the user to lock the cursors in the frequency domain. This is usually done to display the magnitude of the 1× vibration for all six channels.

Once the machine is running stable, the user can click on the "Average and Save" button. This will take eight averages of the spectral data and then save the spectra, time waveforms, and overall values in a job folder on the corporate server. This is critical. Plotting data to paper only "freezes" the data. If the data are needed for future interrogation, the user will not be able to change the frequency viewing span, move cursors, and so on. Saving the raw data allows for this flexibility.

There is also the ability to save the data during a coast-down of the machine. Frequently, the reason for the vibration being slightly higher than desired can be identified by looking at what happens to the magnitudes during a coast-down.

COAST-DOWN DATA

The coast-down plot for this motor identifies the 7,200-cpm peak as being 100% electrically induced (Figure 10.2). Notice that as soon as the power was cut to the motor, the peak was gone! The other important piece of information uncovered in this plot is a natural frequency just below running speed. The 1× peak is identified by the cursor readout and solid dots on the top of the peaks in the spectra. As the motor coasted down, without power, the vibration peaked at 3,314 rpm. Another natural frequency is identified at 750 rpm. The natural frequency at 3,314 rpm appears to be removed from the 1× vibration just enough to not cause an issue because the vibration while the motor was at speed was just over 0.06 inch/s velocity.

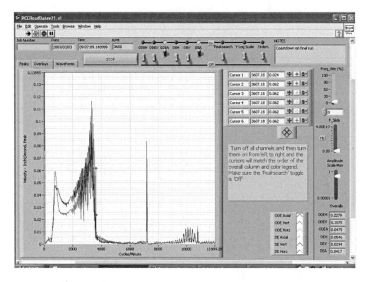

Figure 10.2   Coast-down plot captured on motor.

Note: The overall values have no significance for a peak-averaged coast-down plot.

TACHOMETER AND PHASE INFORMATION

A seventh channel was added for a tachometer input. This additional information will allow the test stand technician to identify whether the vibration at running speed at each end of the machine is in phase or not. These data are critical for differentiating between a first and second rigid-body mode. The information also allows the technician to balance at the test stand (Figure 10.3). There are many balancing programs that will run on a smart phone. All the technician needs is the magnitude and phase readings for the original and trial runs.

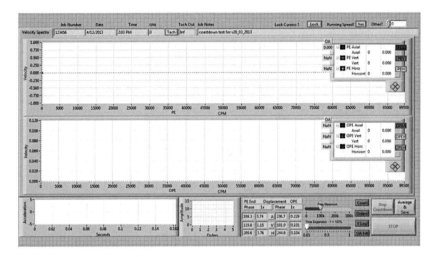

Figure 10.3 Vibration test stand with phase channel.

Additional software is used for postprocessing the saved data. The software can be provided to customers to download and view their own data.

VIEWING THE FAST FOURIER TRANSFORM (FFT)

The data recalled for this example were taken with a tachometer input (v20 datafile). The motor was a DC motor operating at 2,680 rpm, as identified in the "RPM" field at the top of the screen (Figure 10.4). The user will be presented with the "Job Number," "Date," and "Time" the data were taken; "RPM" from the saved data file; and "NOTES" from the test stand user. The default wake-up state is all six velocity spectra overlaid in one display. The bottom of the screen indicates the four key data constraints to ensure that the data were saved properly. The right side of the screen has the options the user might want to use for interrogating the data. All displays are now auto-scaled.

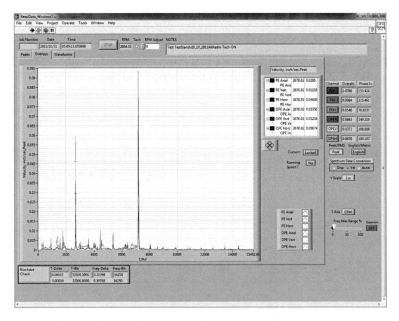

Figure 10.4   Initial screen after data have been recalled.

The "Phase 1X" column has the phase relationship of the running-speed peak for all six axes. The user can interrogate the phase relationship of any two axes on the motor frame.

Note: The lower left box contains the "Block Size" check of the recalled data to ensure that the file has been recalled correctly.

PEAK SEARCH LIST

The user can click on the "Peaks" tab to bring up one of the traces in a peak search plot. This will identify the 1× revolutions per minute peak in the tabular list at the right side of the plot (Figure 10.5). The "RPM Adjust" is then used to adjust the revolutions per minute until the "Order" column indicates "1.0000" (Figure 10.6). Notice that the revolutions per minute are now more accurate with a reading of 1798.37 rpm.

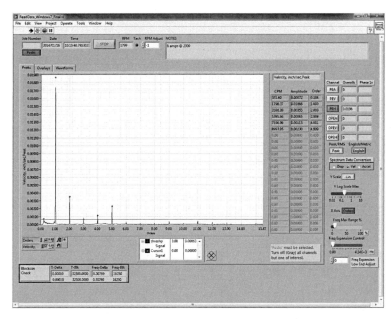

Figure 10.5   Peak search plot with six peaks identified on the pulley-end horizontal axis.

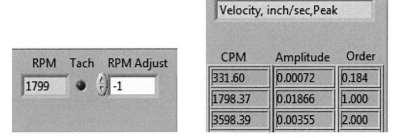

Figure 10.6   "RPM Adjust" with close-up of the tabular output.

Note: The most used controls for modifying the viewed data are defaulted at the startup of the application.

TIME AND FREQUENCY EXPANSION CONTROLS

This is an example of expanding the frequency axis and then using the "Freq Expansion Control" to scroll through the FFT data (Figure 10.7). In this case, the minimum frequency value is 6.56 orders of running speed, or 17,580 cpm. This might be performed to scrutinize a frequency that may be related to ball bearings. The expanded window, once created, can be used to scroll through the entire 300,000-cpm range.

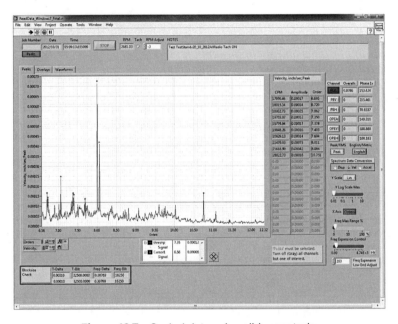

Figure 10.7   Scaled data using slider controls.

MULTIPLE-SPECTRA OVERLAY VIEW

This particular display has the *x*-axis set to "Orders" (Figure 10.8). Notice that the 1× running-speed peak in the data resides above the "1" on the *x*-axis. Changing the "X Axis" button to "CPM" will change the display to "Cycles per Minute."

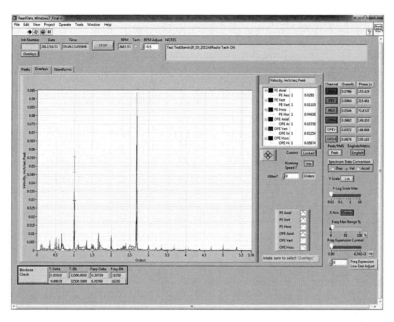

Figure 10.8   Scaled data—overlay of all velocity spectra.

The "Spectrum Data Conversion" selections give the user the ability to view the FFT data in displacement, velocity, acceleration, or demodulated spectra.. In addition, just above this control are additional buttons to choose "Peak" or "RMS" scaling, as well as "English" or "Metric" units.

Note: There is a button just below the cursor readouts labeled "Cursors" with a selection button that reads "Locked." The wake-up default state locks all six spectral cursors on the user input running-speed peak. Below this is another button labeled "Running Speed?"

Again, this is the default. The user can click this button from "Yes" to "No." The user can then type in any frequency he or she might want all cursors to move to, for example 7,200 cpm. The cursors can be easily moved back to running speed by clicking that button back to "Yes."

Because the 7,200-cpm peak is not easily identified as an order of running speed on a DC machine, the user can click off "X Axis" from orders to "CPM" and then type "7200" into the "Other?" field. All six cursors now will move to 7,200 cpm (Figure 10.9).

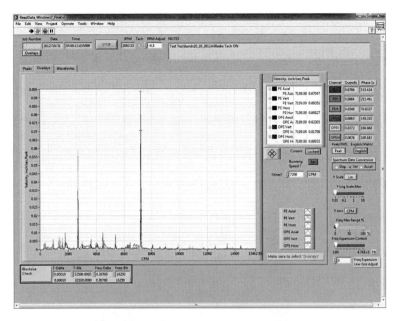

Figure 10.9 All cursors moved to 7,200 cpm.

The locked cursors can be used to jump through the orders of running speed. When the "X Axis" is in "Orders," the user can just increment the "Other?" selection button to move from the first order of running speed to the second, third, and so on, and all the cursors will move together.

The user can select the FFT data type (Figure 10.10). Notice the small selector in the upper middle of the right-hand column. The user just clicks on "Disp," "Vel," or "Accel." The amplitude axis will be relabeled "Displacement, mils P-P." Also, all six cursor values and the "Overalls" will be changed to displacement. The "Peak/RMS" button is disabled for displacement because all readings for displacement are peak to peak ("P-P").

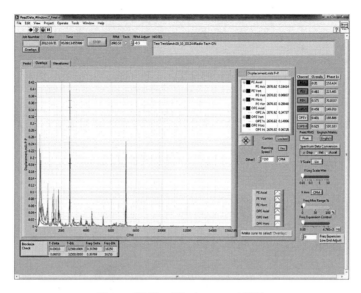

Figure 10.10 Displacement FFT.

Here the "Metric" button is active. Note that the y-axis has been relabeled to "Displacement, u P-P." The "u" symbol stands for microns, the standard metric measurement unit for displacement (Figure 10.11).

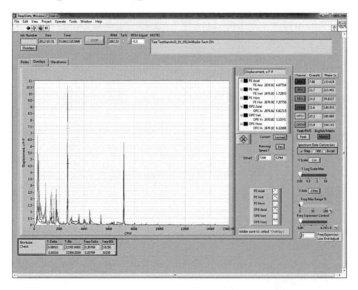

Figure 10.11 Displacement FFT with metric units active.

The "Y Scale" control turns on or off a logarithmic scale on the vertical axis. This can be very useful in identifying resonance issues in the vibration data. The data in Figure 10.12 indicate that there is a natural-frequency response just below the running-speed peak (1,800 rpm). It is identified by the width at the base of the 1× running-speed peak. Also, there is a broader response (more damping) below the 3× running-speed peak and another between the 4× and 5× running-speed peaks.

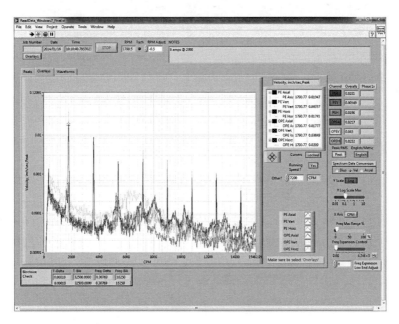

Figure 10.12   Example of *y* log scaled data.

VIEWING TIME WAVEFORMS

The data collected for time waveforms include 32,000 points for each channel (Figure 10.13).

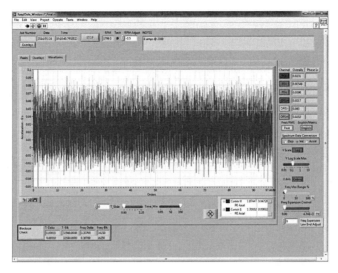

Figure 10.13   Overlay of all acceleration-time waveforms.

The sliders at the bottom of the "Waveforms" tab work exactly like those on the "Overlays" tab (Figure 10.14). The user can expand the 3.25 seconds of time data and then scroll through the entire 3.25 seconds of data with the "Expand" window.

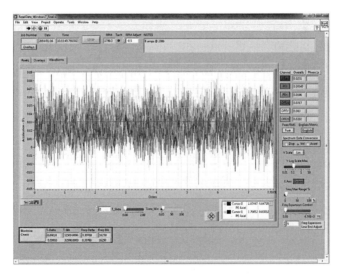

Figure 10.14   Expanded view of all acceleration-time waveforms.

Note: The "Y Scale" is linear. There is no log scaling of time data.

VIEWING PEAK SEARCH DISPLAY

These data illustrate a vibration signature where the user might want to know what is modulating near the third order of running speed (Figure 10.15).

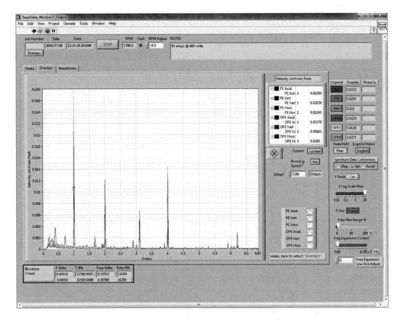

Figure 10.15   Velocity data with unidentified sidebands.

The "Peaks" tab works on only one channel at a time. The user clicks on the "Peaks" tab and also clicks the "Peaks" button above the tab. Then the user turns off all channels but one. In this case, the channel being interrogated is "OPEV."

Here the user has adjusted the "RPM Adjust" to identify the 1× running speed as 1799.5 rpm. Notice that there is a predominant peak near the third order of running but not exactly 3×.

The table in Figure 10.16 lists the peaks cycles per minute (cpm), amplitude in selected user units, and the order of running speed. In this example, there is a peak at 3.103 orders. That happens to be the ball pass frequency of the outer race for the ball bearing in this motor. This example was incoming, and the vibration levels are very low.

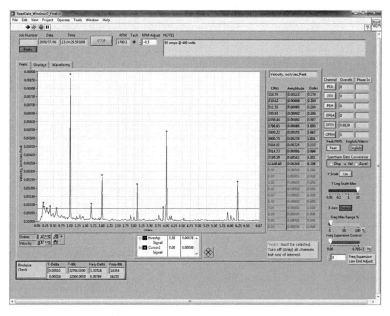

Figure 10.16   Peak search of one channel.

Figure 10.17 is an example of more peaks than the list can hold. If the user were interested in identifying more peaks in the 40–68 orders region, that can be accomplished using the "Freq Expansion Control."

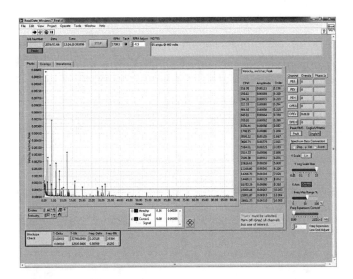

Figure 10.17   An example of more peaks than the list can hold.

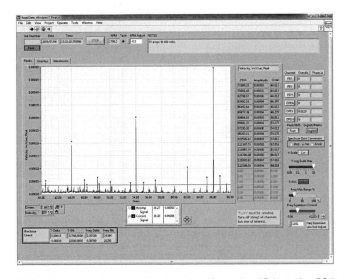

Figure 10.18   Example of expanding the view from the 40th to the 68th orders.

Now the user can identify peaks in a very small range, and all the peak values fit in the list. Notice the position of the "Frequency Expansion Control" slide on the right.

COAST-DOWN EXAMPLE

The user can recall the coast-down data from the job folder. The plot in Figure 10.19 contains coast-down information from all six axes. The "OPEV" and "OPEH" axes, because this is a vertical motor, represent the in-line and perpendicular radial readings from the top of the motor. They also illustrate that the top of the motor is vibrating at the same magnitude in both orthogonal directions during the coast-down. This would present an orbit that is a perfect circle. The "PEV" and "PEH" axes also indicate that they are moving similarly. Note how low the axial vibrations are at both ends.

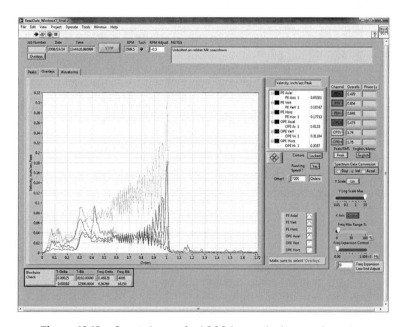

Figure 10.19   Coast-down of a 1,200-hp vertical two-pole motor.

The "Y Log Scale" is used to allow the user to see at one time all information contained in the higher magnitudes, as well as other energies that have very low amplitudes (Figure 10.20). In this case, it is very helpful in identifying a natural frequency that is in resonance just below the running speed. *Coast-down captures* or, as they are also called, *peak hold–averaged plots*, capture every magnitude during a run-down or run-up and save those data in each bin of the FFT. If a higher value than the stored value is encountered, the higher value overwrites the original value.

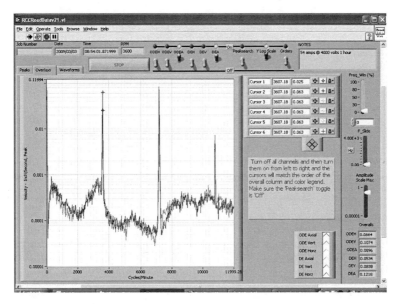

Figure 10.20   Normal averaged velocity spectra with "Y Log Scaling" turned on.

Figure 10.21 shows the same motor that was used to illustrate the difference between a linear and logarithmic display. There are two very important items captured by these data:

1. The vibration amplitude at 7,200 rpm (2× line frequency) dropped immediately on cutting the power, identifying it as electrically induced vibration.
2. The 1× vibration amplitude increased as the motor slowed down through 3,314 rpm. The amplitude then dropped off, increasing again as motor speed passed through 750 rpm. This identifies two resonant frequencies. The one at 750 rpm was due to operating the motor on isolation pads. The one at 3,314 rpm identified a rigid-body mode of the rotor.

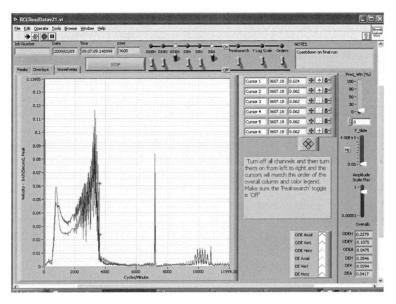

Figure 10.21 Coast-down of two-pole motor (read data view).

Important: The overall data and the phase data are meaningless for a coast-down plot.

Figure 10.22 is an illustration of the data available from a coast-down on a two-pole motor. The 7,200-cpm peak drops drastically on "Power Off," indicating that it is electrically induced. The motor 1× vibration drops to a minimum peak at approximately 3,167 rpm and then climbs to a high vibration magnitude at 2,104 rpm, indicating a natural-frequency response. The user can read down the magnitude column to find the maximum and then identify its frequency or order of running speed.

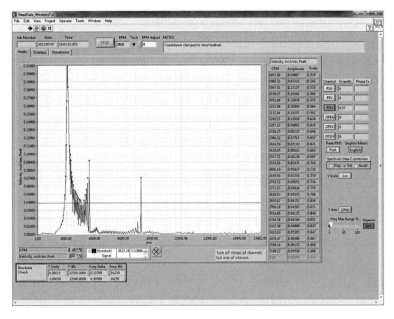

Figure 10.22 Peaks identification on coast-down data.

This version of the test stand system was written to allow for the additional collection of phase data (Figure 10.23). All the features that exist in the basic test stand system are included. The following are additions: Next to the "RPM" field at the top is a green "Tach" button and a readout of the "Tach Out" signal. If the user has attached a tachometer to channel 7 of the NI 2345 front end, this button is used to allow that tachometer signal to be read by the software. There is also a time window at the bottom of the screen that will show the pulse signal from the tachometer output.

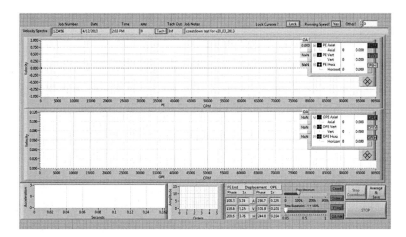

Figure 10.23   Test Stand, version 20 (Windows 7).

The field next to the "Tach Out" time window is the information most useful to the user. The readings from the triaxial accelerometers have been output as a displacement reading in mils peak to peak. Next to each of these values is the phase in degrees. This information can be useful in determining the motion of the bearings end to end, horizontal to vertical, and so on.

Note: The placement of the accelerometers on the motor frame is very important for the phase readings. If the accelerometers are placed on the frame with the cables opposite each other, then the axial phase readings will be 180 degrees out from the actual phase relationship.

THINGS TO TAKE AWAY FROM DISCUSSIONS ON A TEST STAND VIBRATION ACQUISITION SYSTEM

- All channels should be acquired simultaneously, including all six directions normally tested on rotating equipment.
- The system should be able to acquire the peak hold–averaged spectrum on all six directions simultaneously.
- The data acquired should be velocity spectra and acceleration-time waveforms.
- The data should be saved in the job folder and archived in the corporate file system.

- The data should be retrievable and the spectra viewed in all four vibration types: displacement, velocity, acceleration, and demodulated acceleration.
- Overall values for the spectra also should be viewable for each of the vibration types.
- Cursors should be easily placed at the peaks of various frequencies, such as the harmonics of running speed or any other frequency of interest—for example, the silicon-controlled rectifier (SCR) firing rate for DC motors.
- A peak search mode should identify the peaks not only in frequency but also in order of running speed.
- Expand the spectra and time waveforms for ease of viewing data in a given range.
- The system must be easy to use. Technicians assigned to the test stand must be able to understand the system because they will be using the system on an almost daily basis.
- Phase information from the 1× order of running speed should be acquired so that the system can be used in conjunction with a computer balancing program to balance at the test stand without tying up a portable data collector, which should be out in the field.
- The system should have the ability to dump any screen to the office printer for inclusion in the on-site job folder.
- Viewing the data should include "CPM" and "Orders" for the x-axis and "Y lin" and "Y log" scales for the magnitudes.

Glossary of Vibration Terms

Acceleration. The time rate of change of velocity. Typical units are ft/s/s, meters/s/s, and g's (1g = 32.17 ft/s/s = 9.81 m/s/s). Acceleration measurements are usually made with accelerometers.

Accelerometer. Transducer whose output is directly proportional to acceleration. Most commonly use piezoelectric crystals to produce output.

Aliasing. A phenomenon which can occur whenever a signal is not sampled at greater than twice the maximum frequency component. Causes high frequency signals to appear at low frequencies. Aliasing is avoided by filtering out signals greater than ½ the sample rate.

Alignment. A condition whereby the axes of machine components are either coincident, parallel or perpendicular, according to design requirements.

Amplification Factor (Synchronous). A measure of the susceptibility of a rotor to vibration amplitude when rotational speed is equal to the rotor natural frequency (implies a flexible rotor). For imbalance type excitation, synchronous amplification factor is calculated by dividing the amplitude value at the resonant peak by the amplitude value at a speed well above resonance (as determined from a plot of synchronous response vs. rpm).

Amplitude. The magnitude of dynamic motion or vibration. Amplitude is expressed in terms of peak-to peak, zero-to-peak, or rms. For pure sine waves only, these are related as follows: rms = 0.707 times zero-to peak; peak-to-peak = 2 times zero-to-peak. DSAs generally read rms for spectral components, and peak for time domain components.

Anti-Aliasing Filter. A low-pass filter designed to filter out frequencies higher than 1/2 the sample rate in order to prevent aliasing.

Anti-Friction Bearing. *See* Rolling Element Bearing.

Asymetrical Support. Rotor support system that does not provide uniform restraint in all radial directions. This is typical for most heavy industrial machinery where stiffness in one plane may be substantially different than stiffness in the perpendicular plane. Occurs in bearings by design, or from preloads such as gravity or misalignment.

Asynchronous. Vibration components that are not related to rotating speed (also referred to as nonsynchronous).

Attitude Angle (Steady-State). The angle between the direction of steady-state preload through the bearing centerline, and a line drawn between the shaft centerline and the bearing centerline. (Applies to fluid film bearings.)

Auto Spectrum (Power Spectrum). DSA spectrum display whose magnitude represents the power at each frequency, and which has no phase. RMS averaging produces an auto spectrum.

Averaging. In a DSA, digitally averaging several measurements to improve accuracy or to reduce the level of asynchronous components. Refer to definitions of RMS, time, and peak-hold averaging.

Axial. In the same direction as the shaft centerline.

Axial Position. The average position, or change in position, of a rotor in the axial direction with respect to some fixed reference position. Ideally, the reference is a known position within the thrust bearing axial clearance or float zone, and the measurement is made with a displacement transducer observing the thrust collar.

Balanced Condition. For rotating machinery, a condition where the shaft geometric centerline coincides with the mass centerline.

Balancing. A procedure for adjusting the radial mass distribution of a rotor so that the mass centerline approaches the rotor geometric centerline.

Balancing Resonance Speed(s). A rotative speed that corresponds to a natural resonance frequency.

Band-Pass Filter. A filter with a single transmission band extending from lower to upper cutoff frequencies. The width of the band is determined by the separation of frequencies at which amplitude is attenuated by 3 dB (0.707).

Bandwidth. The spacing between frequencies at which a band-pass filter attenuates the signal by 3 dB. In a DSA, measurement bandwidth is equal to [(frequency span)/ (number of filters) × (window factor)]. Window factors are: 1 for uniform, 1.5 for Hanning, and 3.63 for flat top.

Baseline Spectrum. A vibration spectrum taken when a machine is in good operating condition; used as a reference for monitoring and analysis.

Blade Passing Frequency. A potential vibration frequency on any bladed machine (turbine, axial compressor, fan, etc.). It is represented by the number of blades times shaft-rotating frequency.

Block Size. The number of samples used in a DSA to compute the fast Fourier transform. Also the number of samples in a DSA time display. Most DSAs use a block size of 1,024. Smaller block size reduces resolution.

Bode. Rectangular coordinate plot of 1× component amplitude and phase (relative to a keyphasor) vs. running speed.

BPFO, BPFI. Common abbreviations for ball pass frequency of defects on outer and inner bearing races, respectively.

Bow. A shaft condition such that the geometric centerline of the shaft is not straight.

Brinneling (False). Impressions made by bearing rolling elements on the bearing race; typically caused by external vibration when the shaft is stationary.

Calibration. A test during which known values of the measured variable are applied to the transducer or readout instrument, and output readings varied or adjusted.

Campbell Diagram. A mathematically constructed diagram used to check for coincidence of vibration sources (i.e., 1 × imbalance, 2 × misalignment) with rotor natural resonances. The form of the diagram is a rectangular plot of resonant frequency (y-axis) vs. excitation frequency (x-axis). Also known as an interference diagram.

Cascade Plot. *See* Spectral Map.

Cavitation. A condition that can occur in liquid handling machinery (e.g., centrifugal pumps) where a system pressure decrease in the suction line and pump inlet lowers fluid pressure and vaporization occurs. The result is mixed flow, which may produce vibration.

Center Frequency. For a bandpass filter, the center of the transmission band.

Charge Amplifier. Amplifier used to convert accelerometer output impedance from high to low, making calibration much less dependent on cable capacitance.

Coherence. The ratio of coherent output power between channels in a dual-channel DSA. An effective means of determining the similarity of vibration at two locations, giving insight into the possibility of cause-and-effect relationships.

Constant Bandwidth Filter. A band-pass filter whose bandwidth is independent of center frequency. The filters simulated digitally in a DSA are constant band width.

Constant Percentage Bandwidth. A band-pass filter whose bandwidth is a constant percentage of center frequency. 1/3 octave filters, including those synthesized in DSAs, are a constant percentage bandwidth.

Critical Machinery. Machines that are critical to a major part of the plant process. These machines are usually unspared.

Critical Speeds. In general, any rotating speed that is associated with high vibration amplitude. Often, the rotor speeds that correspond to natural frequencies of the system.

Critical Speed Map. A rectangular plot of system natural frequency (y-axis) versus bearing or support stiffness (x-axis).

Cross Axis Sensitivity. A measure of off-axis response of velocity and acceleration transducers.

Cycle. One complete sequence of values of a periodic quantity.

Damping. The quality of a mechanical system that restrains the amplitude of motion with each successive cycle. Damping of shaft motion is provided by oil in bearings, seals, etc. The damping process converts mechanical energy to other forms, usually heat.

Damping, Critical. The smallest amount of damping required to return the system to its equilibrium position without oscillation.

Decibels (dB). A logarithmic representation of amplitude ratio, defined as 20 times the base ten logarithm of the ratio of the measured amplitude to a reference. DbV readings, for example, are referenced to 1 volt rms. Db amplitude scales are required to display the full dynamic range of a DSA.

Degrees of Freedom. A phrase used in mechanical vibration to describe the complexity of the system. The number of degrees of freedom is the number of independent variables describing the state of a vibrating system.

Digital Filter. A filter which acts on data after it has been sampled and digitized. Often used in DSAs to provide anti-aliasing protection after internal re-sampling.

Differentiation. Representation in terms of time rate of change. For example, differentiating velocity yields acceleration. In a DSA, differentiation is performed by multiplication by jw, where w is frequency multiplied by 2þ. (Differentiation can also be used to convert displacement to velocity.)

Discrete Fourier Transform. A procedure for calculating discrete frequency components (filters or lines) from sampled time data. Since the frequency domain result is complex (i.e., real and imaginary components), the number of points is equal to half the number of samples.

Displacement. The change in distance or position of an object relative to a reference.

Displacement Transducer. A transducer whose output is proportional to the distance between it and the measured object (usually the shaft).

DSA. *See* Dynamic Signal Analyzer.

Dual Probe. A transducer set consisting of displacement and velocity transducers. Combines measurement of shaft motion relative to the displacement transducer with velocity of the displacement transducer to produce absolute motion of the shaft.

Dual Voting. Concept where two independent inputs are required before action (usually machine shutdown) is taken. Most often used with axial position measurements, where failure of a single transducer might lead to an unnecessary shutdown.

Dynamic Motion. Vibratory motion of a rotor system caused by mechanisms that are active only when the rotor is turning at speeds above slow roll speed.

Dynamic Signal Analyzer (DSA). Vibration analyzer that uses digital signal processing and the fast Fourier transform to display vibration frequency components. DSAs also display the time domain and phase spectrum, and can usually be interfaced to a computer.

Eccentricity, Mechanical. The variation of the outer diameter of a shaft surface when referenced to the true geometric centerline of the shaft. Out-of-roundness.

Eccentricity Ratio. The vector difference between the bearing centerline and the average steady-state journal centerline.

Eddy Current. Electrical current which is generated (and dissipated) in a conductive material in the presence of an electromagnetic field.

Electrical Runout. An error signal that occurs in eddy current displacement measurements when shaft surface conductivity varies.

Engineering Units. In a DSA, refers to units that are calibrated by the user (e.g., in/s, g's).

External Sampling. In a DSA, refers to control of data sampling by a multiplied tachometer signal. Provides a stationary display of vibration with changing speed.

Fast Fourier Transform (FFT). A computer (or microprocessor) procedure for calculating discrete frequency components from sampled time data. A special case of the discrete Fourier transform where the number of samples is constrained to a power of 2.

Filter. Electronic circuitry designed to pass or reject a specific frequency band.

Finite Element Modeling. A computer aided design technique for predicting the dynamic behavior of a mechanical system prior to construction. Modeling can be used, for example, to predict the natural frequencies of a flexible rotor.

Flat Top Filter. DSA window function that provides the best amplitude accuracy for measuring discrete frequency components.

Fluid-Film Bearing. A bearing that supports the shaft on a thin film of oil. The fluid-film layer may be generated by journal rotation (hydrodynamic bearing), or by externally applied pressure (hydrostatic bearing).

Forced Vibration. The oscillation of a system under the action of a forcing function. Typically forced vibration occurs at the frequency of the exciting force.

Free Vibration. Vibration of a mechanical system following an initial force—typically at one or more natural frequencies.

Frequency. The repetition rate of a periodic event, usually expressed in cycles per second (Hz), revolutions per minute (rpm), or multiples of a rotational speed (orders). Orders are commonly referred to as 1× for rotational speed, 2× for twice rotational speed, etc.

Frequency Response. The amplitude and phase response characteristics of a system.

G. The value of acceleration produced by the force of gravity.

Gear Mesh Frequency. A potential vibration frequency on any machine that contains gears; equal to the number of teeth multiplied by the rotational frequency of the gear.

Hanning Window. DSA window function that provides better frequency resolution than the flat top window, but with reduced amplitude accuracy.

Harmonic. Frequency component at a frequency that is an integer multiple of the fundamental frequency.

Heavy Spot. The angular location of the imbalance vector at a specific lateral location on a shaft. The heavy spot typically does not change with rotational speed.

Hertz (Hz). The unit of frequency represented by cycles per second.

High Spot. The angular location on the shaft directly under the vibration transducer at the point of closest proximity. The high spot can move with changes in shaft dynamics (e.g., from changes in speed).

High-Pass Filter. A filter with a transmission band starting at a lower cutoff frequency and extending to (theoretically) infinite frequency.

Hysteresis. Non-uniqueness in the relationship between two variables as a parameter increases or decreases. Also called deadband, or that portion of a system's response where a change in input does not produce a change in output.

Imbalance. Unequal radial weight distribution on a rotor system; a shaft condition such that the mass and shaft geometric centerlines do not coincide.

Impact Test. Response test where the broad frequency range produced by an impact is used as the stimulus. Sometimes referred to as a bump test.

Impedance, Mechanical. The mechanical properties of a machine system (mass, stiffness, damping) that determine the response to periodic forcing functions.

Influence Coefficients. Mathematical coefficients that describe the influence of system loading on system deflection.

Integration. A process producing a result that, when differentiated, yields the original quantity. Integration of acceleration, for example, yields velocity. Integration is performed in a DSA by dividing by jw, where w is frequency multiplied by 2þ. (Integration is also used to convert velocity to displacement).

Journal. Specific portions of the shaft surface from which rotor applied loads are transmitted to bearing supports.

Keyphasor. A signal used in rotating machinery measurements, generated by a transducer observing a once-per-revolution event. The keyphasor signal is used in phase measurements for analysis and balancing. (Keyphasor is a Bently Nevada trade name.)

Lateral Location. The definition of various points along the shaft axis of rotation.

Lateral Vibration. *See* Radial Vibration.

Leakage. In DSAs, a result of finite time record length that results in smearing of frequency components. Its effects are greatly reduced by the use of weighted window functions such as flat top and Hanning.

Linearity. The response characteristics of a linear system remain constant with input level. That is, if the response to input a is A, and the response to input b is B, then the response of a linear system to input (a + b) will be (A + B). An example of a non-linear system is one whose response is limited by mechanical stop, such as occurs when a bearing mount is loose.

Lines. Common term used to describe the filters of a DSA (e.g., 400 line analyzer).

Linear Averaging. *See* Time Averaging.

Low-Pass Filter. A filter whose transmission band extends from dc to an upper cut-off frequency.

Mechanical Run-out. An error in measuring the position of the shaft centerline with a displacement probe that is caused by out-of-roundness and surface imperfections.

Micrometer (MICRON). One millionth (.000001) of a meter. (1 micron = 1 × E-6 meters = 0.04 mils.)

MIL. One thousandth (0.001) of an inch. (1 mil = 25.4 microns.)

Modal Analysis. The process of breaking complex vibration into its component modes of vibration, very much like frequency domain analysis breaks vibration down to component frequencies.

Mode Shape. The resultant deflected shape of a rotor at a specific rotational speed to an applied forcing function. A three-dimensional presentation of rotor lateral deflection along the shaft axis.

Modulation, Amplitude (AM). The process where the amplitude of a signal is varied as a function of the instantaneous value of another signal. The first signal is called the carrier, and the second signal is called the modulating signal. Amplitude modulation produces a component at the carrier frequency, with adjacent components (sidebands) at the frequency of the modulating signal.

Modulation, Frequency (FM). The process where the frequency of the carrier is determined by the amplitude of the modulating signal. Frequency modulation produces a component at the carrier frequency, with adjacent components (sidebands) at the frequency of the modulating signal.

Natural Frequency. The frequency of free vibration of a system. The frequency at which an undamped system with a single degree of freedom will oscillate upon momentary displacement from its rest position.

Nodal Point. A point of minimum shaft deflection in a specific mode shape. May readily change location along the shaft axis due to changes in residual imbalance or other forcing function, or change in restraint such as increased bearing clearance.

Noise. Any component of a transducer output signal that does not represent the variable intended to be measured.

Nyquist Criterion. Requirement that a sampled system sample at a frequency greater than twice the highest frequency to be measured.

Nyquist Plot. A plot of real versus imaginary spectral components that is often used in servo analysis. Should not be confused with a polar plot of amplitude and phase of 1× vibration.

Octave. The interval between two frequencies with a ratio of 2 to 1.

Oil Whirl/Whip. An unstable free vibration whereby a fluid-film bearing has insufficient unit loading. Under this condition, the shaft centerline dynamic motion is usually circular in the direction of rotation. Oil whirl occurs at the oil flow velocity within the bearing, usually 40 to 49% of shaft speed. Oil whip occurs when the whirl frequency coincides with (and becomes locked to) a shaft resonant frequency. (Oil whirl and whip can occur in any case where fluid is between two cylindrical surfaces.)

Orbit. The path of the shaft centerline motion during rotation. The orbit is observed with an oscilloscope connected to x- and y-axis displacement transducers. Some dual-channel DSAs also have the ability to display orbits.

Oscillator-Demodulator. A signal conditioning device that sends a radio frequency signal to an eddy-current displacement probe, demodulates the probe output, and provides output signals proportional to both the average and dynamic gap distances. (Also referred to as Proximitor, a Bently Nevada trade name.)

Peak Hold. In a DSA, a type of averaging that holds the peak signal level for each frequency component.

Period. The time required for a complete oscillation or for a single cycle of events. The reciprocal of frequency.

Phase. A measurement of the timing relationship between two signals, or between a specific vibration event and a keyphasor pulse.

Piezoelectric. Any material that provides a conversion between mechanical and electrical energy. For a piezoelectric crystal, if mechanical stresses are applied on two opposite faces, electrical charges appear on some other pair of faces.

Polar Plot. Polar coordinate representation of the locus of the 1× vector at a specific lateral shaft location, with the shaft rotational speed as a parameter.

Power Spectrum. *See* Auto Spectrum.

Preload, Bearing. The dimensionless quantity that is typically expressed as a number from zero to one where a preload of zero indicates no bearing load upon the shaft, and one indicates the maximum preload (i.e., line contact between shaft and bearing).

Preload, External. Any of several mechanisms that can externally load a bearing. This includes "soft" preloads such as process fluids or gravitational forces, as well as "hard" preloads from gear contact forces, misalignment, rubs, etc.

Radial. Direction perpendicular to the shaft centerline.

Radial Position. The average location, relative to the radial bearing centerline, of the shaft dynamic motion.

Radial Vibration. Shaft dynamic motion or casing vibration that is in a direction perpendicular to the shaft centerline.

Real-Time Analyzer. *See* Dynamic Signal Analyzer.

Real-Time Rate. For a DSA, the broadest frequency span at which data is sampled continuously. Real-time rate is mostly dependent on FFT processing speed.

Rectangular Window. *See* Uniform Window.

Relative Motion. Vibration measured relative to a chosen reference. Displacement transducers generally measure shaft motion relative to the transducer mounting.

Repeatability. The ability of a transducer or readout instrument to reproduce readings when the same input is applied repeatedly.

Resolution. The smallest change in stimulus that will produce a detectable change in the instrument output.

Resonance. The condition of vibration amplitude and phase change response caused by a corresponding system sensitivity to a particular forcing frequency. A resonance is typically identified by a substantial amplitude increase, and related phase shift.

Rolling Element Bearing. Bearing whose low friction qualities derive from rolling elements (balls or rollers), with little lubrication.

Root Mean Square (rms). Square root of the arithmetical average of a set of squared instantaneous values. DSAs perform rms averaging digitally on successive vibration spectra.

Rotor, Flexible. A rotor that operates close enough to, or beyond its first bending critical speed for dynamic effects to influence rotor deformations. Rotors that cannot be classified as rigid rotors are considered to be flexible rotors.

Rotor, Rigid. A rotor that operates substantially below its first bending critical speed. A rigid rotor can be brought into, and will remain in, a state of satisfactory balance at all operating speeds when balanced on any two arbitrarily selected correction planes.

RPM Spectral Map. A spectral map of vibration spectra versus rpm.

Runout Compensation. Electronic correction of a transducer output signal for the error resulting from slow roll runout.

Seismic. Refers to an inertially referenced measurement or a measurement relative to free space.

Seismic Transducer. A transducer that is mounted on the case or housing of a machine and measures casing vibration relative to free space. Accelerometers and velocity transducers are seismic.

Signal Conditioner. A device placed between a signal source and a readout instrument to change the signal. Examples: attenuators, preamplifiers, charge amplifiers.

Signature. Term usually applied to the vibration frequency spectrum which is distinctive and special to a machine or component, system, or subsystem at a specific point in time, under specific machine operating conditions, etc. Used for historical comparisons of mechanical conditions over the operating life of the machine.

Slow Roll Speed. Low rotative speed at which dynamic motion effects from forces such as imbalance are negligible.

Spectral Map. A three-dimensional plot of the vibration amplitude spectrum versus another variable, usually time or rpm.

Spectrum Analyzer. An instrument that displays the frequency spectrum of an input signal.

Stiffness. The spring-like quality of mechanical and hydraulic elements to elasticity deform under load.

Strain. The physical deformation, deflection, or change in length resulting from stress (force per unit area).

Subharmonic. Sinusoidal quantity of a frequency that is an integral submultiple of a fundamental frequency.

Subsynchronous. Component(s) of a vibration signal that has a frequency less than shaft rotative frequency.

Synchronous Sampling. In a DSA, it refers to the control of the effective sampling rate of data, which includes the processes of external sampling and computed resampling used in order tracking.

Time Averaging. In a DSA, averaging of time records that results in a reduction of asynchronous components.

Time Record. In a DSA, the sampled time data converted to the frequency domain by the FFT. Most DSAs use a time record of 1,024 samples.

Torsional Vibration. Amplitude modulation of torque measured in degrees peak-to-peak referenced to the axis of shaft rotation.

Tracking Filter. A low-pass or band-pass filter that automatically tracks the input signal. A tracking filter is usually required for aliasing protection when data sampling is controlled externally.

Transducer. A device for translating the magnitude of one quantity into another quantity.

Transient Vibration. Temporarily sustained vibration of a mechanical system. It may consist of forced or free vibration or both. Typically this is associated with changes in machine operating condition such as speed, load, etc.

Transverse Sensitivity. *See* Cross-Axis Sensitivity.

Trigger. Any event that can be used as a timing reference. In a DSA, a trigger can be used to initiate a measurement.

Unbalance. *See* Imbalance.

Uniform Window. In a DSA, a window function with uniform weighting across the time record. This window does not protect against leakage, and should be used only with transient signals contained completely within the time record.

Vector. A quantity that has both magnitude and direction (phase).

Waterfall Plot. *See* Spectral Map.

REFERENCE: Hewlett Packard Dynamic Signal Analyzer Applications; Effective Machinery Measurements Using Dynamic Signal Analyzers, Applications Notes 243–1; October 1991.

Machinery Vibration Diagnostic Guide

Machinery Vibration Diagnostic Guide

INTEGRATED
POWER SERVICES

Probable Source	Predominant Frequency	Dominant Axis	Phase Angle Relationship	Amplitude	FFT Characteristics	Comments
Unbalance						
(1) Mass Imbalance	1x Rotor Speed	Radial (Axial is Higher on Overhung Rotors)	1. Force in-phase (<90 degrees) 2. Couple out of phase (> 90 degrees)	Steady	Narrow Band	Rotor looseness or bow due to thermal stresses may change the amplitude and phase with time.
(2) Bent Shaft	1x RPM (2x RPM If Bent at the Coupling	Axial	180 degrees out of phase axially	Steady	Narrow Band	Run-out at rotor mass appears as unbalance. Run-out at coupling appears as misalignment.
(3) Eccentric Electric Motor Rotor	1x RPM, 1x and 2x Line Frequency	Radial	N/A	Steady	Narrow Band	Will fluctuate (beat) in amplitude and phase if an electrical problem also exists.
Misalignment						
(1) Parallel	1x, 2x RPM	Radial	Radials 180 degrees out of phase	Steady	Narrow Band	
(2) Angular	1x, 2x RPM	Axial	Axials 180 degrees out of phase	Steady	Narrow Band	
(3) Combination	1x, 2x RPM	Radial and Axial	Both radials and axials 180 degrees out of phase	Steady	Narrow Band	

Comments - Most alignments will be a combination of both parallel and angular misalignment. Errors are most common in the vertical plane. Long coupling spans, 1x RPM will be higher in amplitude.

INTEGRATED POWER SERVICES

Machinery Vibration Diagnostic Guide

Probable Source	Predominant Frequency	Dominant Axis	Phase Angle Relationship	Amplitude	FFT Characteristics	Comments
Mechanical Looseness						
(1) Bearings, pedestals, etc. (Stationary)	1x,2x,3x predominant but may extend past 10x with low amplitudes	Radial	Varies with type of looseness	Steady	May extend past 10x	Magnitudes will change drastically across the loose connection (for example, base plate).
(2) Impellers, etc. (Rotating)	1x predominant, but may have harmonics up to 10x with low amplitudes	Radial, Axial for overhung rotor	Will change from start to start	Steady while running, but will change from start to start	May extend past 10x	Variance in amplitude or phase may be caused by center of gravity shifts.
Process Causes						
(1) Blade/Vane Pass Frequency	Njumber of Blades X Running Speed	Radial	N/A	Fluctuating	Narrow Band	More than 1 discharge volute will produce harmonics of blade passing frequency.
(2) Cavitation	Random Broadband	Radial	N/A	Fluctuating	Broadband to 120K	Vane pass frequencies may be superimposed.
Resonance						
(1) Rotor Resonance	1x RPM	Radial	Center hung rotor will display 180 degree out of phase axial relationships	Steady	Narrow band	Appears as mass imbalance except amplitudes will grow faster than square of speed increase.
(2) Base Resonance	Appears close to harmonic exciter	Radial or Axial	Depends of mode shape that is excited	Steady	Narrow band peak with wide base (damping dependent)	Mode shapes must be defined before making corrections.

Machinery Vibration Diagnostic Guide

Probable Source	Predominant Frequency	Dominant Axis	Phase Angle Relationship	Amplitude	FFT Characteristics	Comments
DC Motors						
(1) Faulty Fuses	2x Line Frequency	Radial	N/A	Medium	Steady	AC should be rectified and not end up at the motor.
(2) High torsional vibration	Sidebands around 1xrpm	Radial	N/A	High	Steady	Widen Speed Control Window
(3) Bad SCR	21,600 cpm	Radial	N/A	Grows	Narrow Band	Replace Faulty SCRs
(4) High torsional vibration	Sidebands	Radial	N/A	High	Sidebands	Check Comparitor Inputs. Usually faulty tach.
Drive Belts						
(1) Mismatched, worn, or stretched Also applies to adjustable sheave applications	Many multiples of belt frequency but 2x belt frequency usually dominant	Radial, especially high in line with the belts	N/A	May be unsteady and beating if a belt frequency is close to driver or driven speed	Narrow band	See belt frequency calculation below.
(2) Eccentric and or unbalance sheaves	1x shaft speed	Radial	In-Phase	Steady	Narrow band	Replace sheave
(3) Drive belt or sheave face misalignment	1x drive shaft	Axial	In-Phase	Steady	Narrow band	Perform belt alignment
(4) Drive belt resonance	Belt resonance with no relationship to rotation rpm	Radial	N/A	May be unsteady	Plus or minus 20 percent of resonant frequency depending on dampling	Confirm with strobe light and belt excitation techniques. Change belt tension or belt length.

Belt Frequency = 3.14159 * Diameter * RPM/ Belt Length   NOTE - Chose one sheave for the RPM and Diameter

Machinery Vibration Diagnostic Guide

INTEGRATED POWER SERVICES

Probable Source	Predominant Frequency	Dominant Axis	Phase Angle Relationship	Amplitude	FFT Characteristics	Comments
Bearings						
(1) Anti-friction ball or roller	Early stage - Ball pass frequencies	Radial, except higher axial on thrust bearings	N/A	Increases as bearing condition degrades	Narrow Band	See bearing frequency formulas
	Middle stage - Multiple ball pass frequencies and broad band energy at 30K to 60K CPM	Same	N/A	Same	Ball pass frequencies are narrow band, 30K to 60K energy is broad	Schedule replacement
	Late stage - high 1x rpm multiples		N/A	High. May disappear just before destruction	1x narrow band, broad band energy will lessen in amplitude and grow wider in frequency	Schedule replacement
(2) Sleeve	Early stage - sub-harmonics (may only be detectable with non-contact probe)	Radial	Non-contact probe will indicate shaft position and dynamic changes.	Increases as bearing degrades	High baseline energies below 1x, 2x and 3x RPM	Monitoring of rotor position (thrust) via non-contact probes recommended to protect thrust bearing
	Late stage - Will appear as mechanical looseness	Radial	Same	Same	Same 25% to 33% subharmonics	Schedule replacement
(3) Rub	1/2 RPM and multiple harmonics	Radial	N/A	High	1/2 RPM multiple harmonics	Open clearance on the sides of the babbitt
(4) Oil Whirl, Oil Whip	Critical Speed of rotor	Radial	N/A	High	Once 1x RPM passes through critical speed, vibration will remain high at the crit. freq.	Bearing redesign is necessary to prevent the onset of whip (self damping)
(3) Fluting	BPFO sidebands	Radial	N/A	Grows over time	BPFO Sidebands	Read Chapter Five

Fundamental Train Frequency - FTF = $F/2*[1-(B/P*\cos\theta)]$
Ball Pass Frequency Outer Race - BPFO = $N/2*F*[1-(B/P*\cos\theta)]$
Ball Pass Frequency Inner Race - BPFI = $N/2*F*[1+(B/P*\cos\theta)]$
Ball Spin Frequency - BSF = $P/2B*F*[1-(B/P*\cos\theta)^2]$

B = Ball or Roller Diameter
P = Pitch Diameter of Bearing (measured across diameter, ball centers)
N = Number of Balls or Rollers
$\cos\theta$ = Cosine of contact angle

F = 1 for answers in orders of running speed

Deep groove = cos (0) = 1

Machinery Vibration Diagnostic Guide

Probable Source	Predominant Frequency	Dominant Axis	Phase Angle Relationship	Amplitude	FFT Characteristics	Comments
Electrical Faults						
(1) Eccentric Rotor	2x line frequency	Radial	N/A	Steady	Narrow band	Will fluctuate in amplitude and phase if mechanical issues also exist.
(2) Eccentric Air Gap, and Loose Iron (Stator)	2x line frequency	Radial	N/A	Steady	Narrow band	Loose stator due to soft foot. Shim soft foot to support stator properly
(3) Broken Rotor Bars or Voids in Aluminum Cored Rotor	1,2,3,and 4x orders with #poles x slip sidebands	Radial	N/A	Steady	Narrow band	Excite rotor with sufficient current to isolate discontinuities
(4) Low or missing phase	1/3 line frequency	Radial	N/A	Steady	Narrow band	Check incoming phases for voltage and current
(5) Fluting	BPFO sidebands	Raidal	N/A	Grow with time	BPFO sidebands	Read Chapter 5
VFD Issues						
(1) Shorted Devices or Blown Fuses	2x line frequency	Radial	N/A	Steady	Narrow band	Incoming AC voltage should not show up in motor vibration
(2) Speed Control Tuned too hot	1x rpm	Radial	N/A	Fluctuating	Sideband Modulation	De-tune Speed Control. Too tight.
(3) Improper Grounding Fluting	High frequency multiple peaks at BPFO sidebands	Radial	N/A	High	Sidebanding of High Frequency Peak at BPFO	Read Chapter 5
(4) Faulty Solid State Devices	Current Waveform	Radial	N/A	N/A	N/A	Use On-Line Circuit monitor to inspect firing current waveforms

Machinery Vibration Diagnostic Guide

INTEGRATED POWER SERVICES

Probable Source	Predominant Frequency	Dominant Axis	Phase Angle Relationship	Amplitude	FFT Characteristics	Comments
Gears						
(1) Worn Gears and or support bearings	GMF with sidebands of the modulating gear	Radial	N/A	GMF amplitude will grow and number of sidebands will increase	GMF and sidebands	Repairs should happen immediately to prevent gear damage.
(2) Broken tooth	May not be present	Radial	N/A	Acceleration time waveform sharp peak	N/A	Synchronized Time Averaging should be utilized to emphasize the impact of the bad tooth.
(3) Gear misalignment	2x, 3x GMF	Radial	N/A	N/A	2x, 3x GMF	May require line boring of the gearbox.
(4) Assembly or Manufacturing Errors	Hunting Tooth Freq and/or Gear Assembly Phase Frequencies	Radial	N/A	High	Requires High Resolution FFT	Calculate the FHT by finding the common factor from two interfacing gears Also calculate GAPF.
(5) Machining Errors	3x Running Speed	Radial	N/A	N/A	3x RPM Peak	Overloaded cut of gear in a 3 jaw chuck.

NOTE - Trending Oil Analysis is most important for early detection of faults

Answers to Review Questions

CHAPTER 1

1. Large frequency range; yields raw acceleration-time waveform for impact analysis; small size and low cost.
2. Velocity seismic transducer
3. True. The spring mass system of the magnet must be oil damped to prevent damage.
4. Limited frequency range on low and high ends; very heavy, requiring a large magnet or stud mounting.
5. An ICP accelerometer with a built-in integration circuit will yield a velocity output.
6. A charge-sensitive signal
7. Output signal is voltage sensitive, not charge sensitive. Wire movement doesn't change the output signal. Temperature variables on the cable do not change the output signal. Bends in the output wire do not affect the output signal.
8. 4mA to 20mA constant-current source
9. True. The crystal and the mass attached both get smaller as the frequency range increases.
10. Pros: Highly localized for impacting events, even in very small rotating equipment; free additional measurement from any piezoaccelerometer; will isolate an impacting event before acceleration-time waveforms. Cons:

Difficult to trend because of changes induced by load, speed, and attachment variables; user must identify the frequency of the transducer resonance based on attachment transmissibility.

11. Units of demodulated spectra are given as G_{SE} in Emonitor® products because they are calculated from the raw acceleration-time waveform. They can be trended if mounting is rigid and load and speed do not change, but they are not comparable to acceleration, velocity, or displacement units.

12. Any rotating element where the ratio of the weight of the rotor to the weight of the housing is high; any sleeve-bearing machine where immediate shutdown due to a malfunction is critical.

13. Verify that the automatic shutdown feature of the software has been temporarily disabled.

14. The proximity probe has a dynamic range that will accurately measure up to seven times the fundamental speed of the machine (1× running speed × 7).

15. 0.035 inch, or 50% of its total linear measurement distance.

16. The DC output of the proximeter is read with a meter to center the probe within its useful range.

17. 1 V/0.200 mV/mil = 5 mils peak to peak.

18. Resolution. Increase the lines of resolution for more accurate identification of the individual frequencies.

19. Probe-to-shaft surface gap (DC component). Dynamic vibration between the probe and the shaft (AC component riding the DC voltage).

20. False. The mass of your typical industrial electric motor rotor is more than 30% of the total mass.

21. True. Proximity probes are only useful up to 7× running speed (1× running speed × 7).

22. The gap reading will appear to be a very high vibration if the user does not clarify this.

23. The vibration of the running-speed vibration (1×) will increase.

24. Stud mounting.

25. Spot face the surface to ensure full contact with the accelerometer base. Stud must be perfectly perpendicular to the surface to prevent cocking the accelerometer.

CHAPTER 2

1. RMS is the area under the curve: $0.707 \times 1.2 = 0.8484$ V.
2. $1.2 \times 2/0.200 = 12$ mils peak-to-peak displacement.
3. True.
4. False. Root mean squared.
5. True.
6. False. They are also used for acceleration time waveforms.
7. True.
8. Velocity is in inches per second. The missing parameter is π.
9. 386 inches/s^2
10. Low-end filters (high pass). Velocity data filters are on the order of 320 cpm; 162 cpm is used for acceleration spectra.
11. The amplitude addition and subtraction of two closely spaced but separate frequency vibrations. Two motors operating at 3,590 and 3,580 rpm would have a beat frequency of 10 rpm, or 0.167 Hz.
12. The beat frequency is calculated by subtracting the two frequencies. So, as the difference between the two frequencies decreases, so does the beat frequency.
13. $3,585 - 3,580 = 5$ cpm, or 0.083 Hz.
14. Every 12 seconds the beat will hit maximum.
15. 7,200 seconds/60 cpm $= 0.00833$ cpm, or 0.00014 Hz.
16. MOH, MOV, MOA, MIH, MIV, FIH, FIV, FIA, FOH, FOV.
17. Motor outboard in line (with terminal box), motor outboard perpendicular (to terminal box).
18. The axial reading should be taken on the fixed or thrust bearing.
19. Amplitude and frequency.
20. They are affected by temperature and require periodic recalibration.
21. The Nyquist frequency requires the analysis rate to be at least 2.56 times the maximum frequency of interest. For a maximum frequency range of 1,000 Hz, the sampling rate should be at least 2,560 Hz.
22. 50 hp. Trend for safety reasons only.
23. The number of balls or rollers.
24. Velocity.
25. The ICP power must be supplied by the data collector. This is usually in the collection setup.

26. The 1× belt frequency. Belt 1× speed = π × motor 1× × motor pulley diameter/belt length. ***Note:*** Answer will be the same if the user substitutes fan 1× and fan pulley diameter for motor 1× and motor pulley diameter

27. At least six.

28. Identify harmonics of fundamental frequencies. Find the fundamental frequencies first.

29. 1× running speed or the fundamental frequency of the variable-speed drive.

30. Alarm bands are essential to culling machines that require attention. Automated reports can also greatly cut down on time.

31. Spectral alarm bands will perform a statistical analysis of all readings from all machines or all readings from a given measurement point of all machines. This will isolate the outlier.

32. Reduction is downtime. This information will show value for the program and is essential to proving the worth of the program for its continuation. Downtime costs are readily available from the accounting department. Importantly, these numbers will be on the conservative side, and the potential of losses due to a catastrophic failure are then additional proof of program value.

CHAPTER 3

1. Periodic oil analysis. It will find degradation items in a gearbox faster than any other predictive tool.

2. Parabolic:

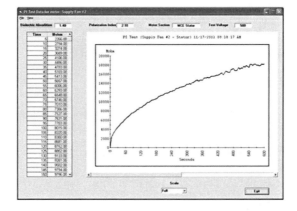

3. The high-frequency energy that has suddenly appeared at the top of this waterfall plot is indicative of a failing ball bearing. The bearings should be replaced ASAP and a root-cause failure analysis should be done to determine why the bearing failed, if premature.

4. Setting of alarm parameters and automated reporting to analyze all recorded data as quickly and as efficiently as possible.

5. False. The data must be reviewed, and any units that require maintenance should be brought to the attention of plant maintenance personnel. Also, corporate accounting should be asked for a monetary figure for savings resulting from the prevented downtime by catching the problematic machinery. These figures will justify to management the benefits of the program.

6. Multiply the number of vanes in the impeller times the running speed of the shaft.

7. This is usually an indicator of a serious change in the mechanical operation of the machine. For example, ball bearings, when they start to fail, have more clearance and therefore may cause a drop in magnitude in a given frequency band.

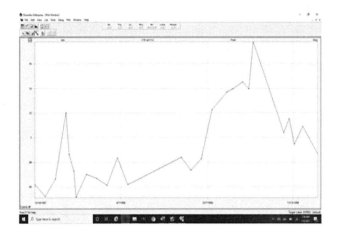

8. False. A two- or three-page summary of any machinery that requires attention is most important for the customer. One plot for each machine to illustrate the problem should be included as an addendum.

9. True.

10. False. Bearings can have any number of reasons why they don't give full life. Trending parameters must be set to view demodulated spectra and acceleration-time waveforms for a quick analysis of bearing conditions.

11. True. Testing a customer's electrical cabinets should only be attempted with a qualified electrician who knows the site rules and the equipment.

12. True. The benefits from time spent viewing electrical cabinets is immeasurable. Overloaded leads, loose connections, and so on can all be isolated quickly.

13. The slinger ring rotates on the shaft, and as it passes through the oil sump, the ring picks up the oil and deposits it onto the bearing journal. Care must be taken not to have too high an oil level because this can slow down the rotation of the ring. For this reason, the oil should be level with the inner diameter of the slinger ring.

14. False. Over-lubrication can cause excess heat in the bearing, reducing bearing life.

15. Usually, regreasing schedules are every 6 months or so. This means that during the last week of that 6-month stretch, there is very little grease left in the load zone. Dividing the amount of grease that would be added every 6 months into 12 increments, the bearings are sure to get new grease every 2 weeks.

16. The tester should be connected to the outgoing leads at the motor control cabinet or the VFD cabinet. This will not only qualify the motor winding but also give an assessment of the entire cable run from the cabinet to the motor terminal box.

17. A PI of 3.0 with a parabolic shape is an indication of a good, solid insulation system.

18. True. An AC hipot should *never* be performed on an in-service motor. The risk of failing the winding is substantial. Only test newly wound machines with AC hipot.

CHAPTER 4

1. A starting point. Data must be taken to eliminate all other issues so that a conclusion can be reached.

2. The total cost of the fix. This will include engineering time, as well as downtime to accomplish the fix.

3. Identify harmonic "families," and then concentrate on what the cause of the fundamental frequency could be. Identify all families and their fundamental cause.

4. 1× running speed. There may be harmonics due to impacting in the bearings, but these will be alleviated when the unbalanced condition is remedied.

5. (Pump keyway length + coupling keyway length)/2.
   (2 + 1)/2 = 3/2 = 1½ inches long.

6. False. Cracks, if the fan is made of weldable steel, should always be repaired first. Dirt must always be removed before balancing any rotor.

7. The rotor has a dynamic (couple) imbalance, meaning that the imbalances on each end of the rotor are not in phase with each other.

8. A ⅛-inch welding rod weighs approximately 20 g, so 40 g would be added by using two full rods.

9. True. Grounding to the fan housing frame will pass damaging currents through the bearings.

10. No. The 1× vibration at 50% of speed is 0.02. Doubling the speed (2×) squared (4×) would equal 0.08 inch/s velocity, which is acceptable. This would be the vibration level at speed if the fan had only an imbalance problem. This fan should be checked for cracks, loose fit to the shaft, and also critical frequency issues.

11. "G" is German for *Geschwindigkeit*, which means "velocity." The 0.4 represents the velocity in millimeters per second, which should be the maximum vibration of the rotor operating without restraints (free-free)—for example, if the rotor were balanced in a soft-bearing machine. The 0.4 also represents the entire rotor, so the number is divided by 2 for each bearing position, or 0.2 mm/s velocity.

12. False. The G specification means that the rotor has met the absolute maximum level of imbalance. The operator should always make a couple more adjustments to achieve a lower imbalance level.

13. The customer must be contacted to ascertain the industrial use of the rotor and the speed at which it normally operates. Then the technician can weigh the rotor and use the Unbalance Tolerance Guide for Rigid Rotors chart. Again, it is always better to achieve the lowest possible imbalance, within reason.

14. False. Multiple-component rotors should be balanced as an assembly, match marked for angular positioning of the components, and disassem-

bled and then reassembled in the machine, paying close attention to the match marks.

15. Check the velocity spectra for the predominant frequency of the vibration. It will be something other than fan 1×. It could be electrically generated by the motor, bad belts, and so on. Find the cause and then attack that problem next.

16. The G specification is based on mass and speed. The wheels, for their mass, turn at slow speed.

17. Response to unbalance due to the weight is linear at different radii. The mass is being placed at a radius that is 3× the trial weight location, so the final mass is 70 g. Because the welding rod will account for 20 g of this, the plate should be cut to weigh 50 g.

18. So he can remove it in the field! High-speed machinery has the potential to operate near or above a first-bending-mode natural frequency. When this occurs, the high spot on the rotor and the heavy spot switch places. This means that the weight added to the fan in the balance machine operating at low speed will be in the exact opposite position for an acceptable balance at speed. The weight, painted with red paint, was removed in the field after the at-speed vibration readings were found to be unacceptable. The result of removing that weight left the fan running with excellent levels.

CHAPTER 5

1. 1½ to 2 hours.

2. The built-in stresses from improperly seating rotor laminations have relieved themselves, and now the rotor is unbalanced.

3. $2 \times 2 \times 0.25$ inch = 1 inch^3; $1 \times 0.283 = 0.283$ lb.
   1 lb = 454 grams. $0.283 * 454 = 128.5$ g.
   128.5 g weight plus 20 g of welding rod = 148.5 g.
   Note: Density of steel is 0.283 lb/inch^2.

4. Force should be 1/10 of rotor weight, or 100 lb.

Force (lbs) =

$$1.77 \times \left(\frac{\text{RPM}}{1{,}000}\right)^2 \times \text{ounce-inches}$$

100 lbs =

$$1.77 \times \left(\frac{3{,}600}{1{,}000}\right)^2 \times \text{ounce-inches}$$

ounce-inches = 4.36

Don't forget that this is 4.36 oz at a 1-inch radius. If the weight radius is 6 inches, then the estimated trial weight is 4.36 oz/6 inches, or 0.73 oz.

5. Trial weight insufficient to generate a force of 10% of the rotor weight.

6. Large changes in trial weight results could be due to the rotor running near a critical frequency. Therefore, the amount of weight necessary to balance the fan will be less than the calculated result. The newbie should make a change from 100 to 200 g and make another run. Remember, the main objective is to get a change in the vibration level and/or phase. Once a change has been registered, the balance programs will work efficiently.

7. False. The four-run method will work well. This method is also very effective for rotors operating near a bending critical frequency because the phase information can be erroneous at critical frequency.

8. False. Steel trial weights should be used at the edge of steel fan wheels and the weight tack welded on. If the fan is made of another material, other methods should be used, such as using the leading edge of the fan blade for trial weight placement. The centrifugal force will keep the trial weight on the fan blade.

9. True. This will result in a rotor that will run much smoother as the bow takes effect, moving back to the original phase angle. Basically, the vibration levels will continue to drop from startup until they reach a very low level as the phase switches 180 degrees. Then, as it continues to come to its final bow, the vibration levels will be very low.

10. Peak averaging will save the maximum value at every spectral line during a coast-down. The technician should start the average and then shut the machine off. As the values are saved, the technician can watch to ensure

that the 2× vibration doesn't write over the 1× vibration values by stopping the average before the 2× vibration peak reaches the initial running speed at the start of the average.

11. Change the *y*-scale viewing axis to a log scale. This will enhance low amplitude values, and natural frequencies will appear as wide-based peaks in the spectrum. They can also tell the technician if the forcing function (the peak) is above or below the natural frequency. This can be very helpful when attempting to correct the problem by answering the question of whether the natural frequency should be pushed higher in frequency or the forcing function moved higher if possible.

12. False. Unbalance can be a major contributor to 1× running-speed vibration, but misalignment, running at resonance, and so on can all amplify 1× vibration levels.

13. True. The actual weight required to trim out high 1× vibration when a rotor is running near a resonant frequency will be less than what the program will call for.

14. Always be prepared to use every tool in your truck. There are more ways to get to a successful result, and just because the standard approach isn't available (i.e., phase readings, constant speed, and so on), don't give up. Think through the process of what information is needed and how to achieve that gathering of data. Then follow through with the process to correct the problem.

CHAPTER 6

1. The technician has forgotten to check the window in the data collector. The normal Hanning window, used for periodic steady-state vibration, is not to be used for triggered impact analysis. The window should be changed to exponential or flat top or none. Make sure to change the window when going back to collection of running vibration data.

2. The tip of the hammer must match the application. The mass of the hammer is also very important based on the size of the structure under test. A very small gear pinion (2–5 lb) requires a very small hammer with no extended weight and a steel tip. A large electric motor (1,000–5,000 hp) requires a very soft-tipped hammer (soft rubber) and an additional 20 lb bolted to the back of the impact surface.

3. These pumps normally have a discharge pipe that extends orthogonally from the pump housing. The supply to the pump is vertical and centered below the pump. The discharge pipe adds stiffness to the pump-motor support. Therefore, the higher of the natural frequencies is usually in the direction of the discharge piping. The lower frequency is orthogonal to the discharge pipe.

4. The motor and pump could have very low vibration levels when tested during manufacturing, but once that system is assembled in the field, there are many items that can cause a slight increase in unbalance, such as the coupling, or slight run-outs of the shafts, coupling, and so on, which also can cause an increase in the 1× vibration level. If this motor-pump assembly is operating very close to a reed frequency (system critical), the vibration levels can be lowered to acceptable limits by trim balancing at the top of the motor. This will reduce the 1× excitation that is causing the motor-pump, while running at resonance, to not meet specification limits.

5. The motor-pump assembly was not out of balance but was operating near reed-frequency critical speeds of the assembly. With the brace in place, those reed frequencies were moved above all running speeds, and the resulting vibration was now well within satisfactory limits. The original vibration level of 22 mils at a pump operating speed has now been lowered to <0.4 mils.

6. False. The natural frequency of the motor end bracket will be different running than at rest. At rest, the stiffness of the rotor will be coupled into the end bracket stiffness and the reading will be higher in frequency.

7. True. As long as the rotor bearing journals are on an oil film, the reading will be successful.

8. The motor, when operating on 60 cpm power, will always have a significant excitation frequency at 2× line frequency no matter the number of poles.

9. The motor bracket was in resonance. The motor shaft was *not* vibrating at 0.6 inch/s, only the bearing housing was. The oil film under the motor journal was not coupling the bracket/bearing axial vibration into the motor shaft.

10. None. The motor ran for 20 years with the same bearings. All bearing motion was axial, under the oil film.

11. Total equivalent support stiffness calculations include *everything*, including what is below the concrete slab. Frequently, underground water systems

wash away the stone fill that supports the concrete. This will lower the stiffness and may result in a natural frequency moving downward and ending up in close proximity to a forcing function of the machine.

12. False. Too large a mass attached to a fan surface can run the risk of tearing the fan material apart. Rule of thumb for balancing out vibration due to a bent shaft: 0.005 inch or less. There are also instances where bent-shaft problems can be lessened by rotating a coupling during alignment. This is done by placing the maximum run-out of each shaft 180 degrees from each other. Again, this will only work for low amounts of run-out.

13. True. The oil wedge will not form properly, and the journal will have a vibration component that is with rotation at approximately 40% of running speed. This vibration will also generate harmonics.

14. The clearances at the sides of the bearing are too tight. The oil is inhibited from getting down inside the sleeve to create the oil film for the journal to ride on. This vibration can be very severe and damaging. The unit must be shut down immediately and the bearing clearances corrected.

15. The oil temperature can be permitted to increase, as long as it is within tolerance, by cutting back on the oil cooling system. Again, this must be done with everything being within system specifications.

16. The vibration phenomenon is known as *oil whip*. This problem is usually corrected by fitting new-design bearings, such as tilt-pad style.

17. False. The bearing clearances will be compromised, and the bearing will fail quickly due to overheating.

18. A motor with ball bearings should never be shipped without "blocking" the shaft. The bounce of the rotor during transit will cause deformations in the bearing races from impacts with the balls/rollers. This is referred to as *brinelling*.

19. True. Movement of the balls and/or rollers while the machine is standing still can smooth over the final grinding marks. However, if there is any deformation to the race surfaces, the bearing should be replaced ASAP.

20. The design engineer performed his calculations based on the motor at rest. During shipping, the motor can be subjected to shocks that will prove to be in excess of the bearing loading capability. All shipped rotating equipment should have all shafts and assemblies blocked.

21. Measure the coupling inside diameter and the shaft. The difference should be 0.0005 inch per inch of shaft diameter.

22. The bearings need to be loaded axially with the addition of wavy-spring washers. The loading must be calculated because the solution may require multiple washers on each bearing, and overloading the bearing must be avoided.

23. Five to seven harmonics of running speed.

24. The laser and prism are on in reverse position. Switch them and redo. *Note:* I learned this the hard way—120 ft up in a crane during a blizzard!

25. False. It is my experience that based on the stiffness of the I-beam rails, the tightening of the foot bolts often can result in pulling the beam up to the foot instead of pulling the foot down to the beam surface.

26. (a) True. A scribe mark should be left on the shaft to indicate where this is.
(b) True. This is critical when aligning the motor—even more important than magnetic center.
(c) True. Reliance Electric motors are fitted with movable shims.
(d) False. The force due to a rotor not operating on magnetic center is very minimal, and therefore it is more important to keep the clearance at the mechanical end stops.

27. Check the specifications for the bolt strength based on lowering the diameter. The equipment manufacturer will have a specification for torquing down the foot bolts. Ensure that the necked-down bolt will still take that torque loading.

28. The inboard feet of *both* the drive and driven equipment are too low. Add 0.100 inch of shims to the inboard feet of both machines and start over.

29. The technician can add 0.0003 inch per inch of spool piece. So, for example, if a two-pole motor has a parallel alignment tolerance of 0.002 inch, the addition allowance would be 0.011 inch for a total of 0.013 inch. Allowance is only on the parallel readings, not the angular readings.

30. False. The technician, if he suspects that this is the predominant problem, can use a laser alignment system to identify the movement of the drive and driven equipment. Mount the laser head on a bracket of the driver, and mount the prism on a bracket mounted to the driven equipment. Zero them before the unit is started and loaded. After everything is up to operating temperature, take the laser readings. These readings will instruct the technician where the starting alignment should be by starting with these offsets. Then, as the machine comes up to temperature, the alignment readings should approach a very good set of readings.

I recently consulted on the alignment of a generator to a wind turbine gearbox. The OEM specification called for 0.095 inch of thermal offset. The offset was calculated into the cold alignment. The final at-temperature alignment was within 0.002 inch.

31. False. Machinery alignments can be affected by thermal expansion of process ductwork, binding support pivots, and so on.

32. False. Belt tension should always be confirmed with a belt tension gauge.

CHAPTER 7

1. Total shims under any one foot should be limited to three or four. Polished-steel plate should be used if required shimming is more than 0.250 inch.

2. A peak hold analysis will identify any natural frequencies that may be in close proximity to forcing functions. This goes for machines in the field as well as motors on the test stand.

3. True. The engineer controls the bending modes by modifying the shaft stiffness and rotor mass. The rigid-body modes are the responsibility of the application design engineers who lay out the foundation for the equipment.

4. The critical speed map will inform the technician where the total support stiffness resides based on the reaction of the rotor as it passes through the first and second rigid-body modes.

5. The one end of the motor has a natural frequency that is only 5% removed from running speed. Note that the vibration levels of the 1× component increase as the motor speed slows from maximum speed.

6. The variable-speed drive needs to have that speed range blocked out. The motor will not stop within that range. It will be adjustable up to within 10% of that speed, and then, if an increase is requested, it will move to a frequency 10% above that speed.

7. 50,000 lb./inch is very low for a machine of this size. After finishing this work, the floor was investigated and found to be just a 3-inch-thick concrete pour on top of corrugated steel sheets. These sheets rested on 3-ft web I-beams, but there was no mechanical connection! It was a floating floor!

8. The motor design engineer can only control the resulting frequency of the first bending mode. The critical speed map identifies this as approximately 20,000 cpm. This is excellent for a two-pole motor, even if the final design includes overspeed on a variable-frequency drive.

9. This result will yield the vertical line for equivalent support stiffness, and this will establish where the rigid-body modes might result in high vibration.

10. False. The design engineers encountered rigid-body modes at test with only the corners supported. For this reason, the entire length of the machine must be shimmed.

11. The energy was due to 2× line-frequency excitation. When the power is cut, any energy due to electrical energy will disappear. There may still be vibration present in this region, but it will be due to mechanical issues, such as 2× mechanical vibration on a two-pole induction motor.

12. Second rigid-body mode. The ends are out of phase with no bending of the rotor shaft.

13. This second rigid-body mode is close to 3× running speed of a two-pole motor. This could be a problem if there is sufficient energy at that harmonic.

14. The plot is usually known as a frequency-response function. The plot contains the natural-frequency magnitudes and phases from an impact analysis. In this case, there are three axes plotted on one plot: horizontal, vertical, and axial.

15. The plot is the result of taking the FRF plot from above and then double integrating the data and taking the reciprocal. This yields the stiffness of the test piece from the impact analysis. The technician can see that everywhere there is a natural-frequency peak in the FRF, there is a valley in the dynamic stiffness plot.

16. A machine with a natural-frequency problem that can't be remedied by changes in stiffness or mass, and damping methods will not yield acceptable vibration limits.

17. The tuned absorber creates two natural frequencies in the system, one lower and one higher than the problematic resonance problem. *Note:* A tuned absorber will not be a practical solution if the machinery operates on a VFD and will run at either of the two new natural frequencies.

18. Cross-talk from power lead to power lead will result in voltage spikes. These voltage spikes will cause turn-to-turn shorts in the motors.

19. Maximum length power lead run.

20. The structure will have visible motion at those frequencies represented by peaks in the FRF. There will also be zero motion at those frequencies represented by the antinodes.

21. There is as much energy above unity as there is below. Draw a line across the FRF at the magnitude line that resembles unity (the same energy measured by the accelerometer as has been imparted into the structure by the hammer blow), and there will be an identical amount of energy above and below that line.

CHAPTER 8

1. False. Any motor with a frame size greater than a 444 can have low-frequency circulating currents that will cause EDM damage.
2. False. This amount of damage can occur in several months.
3. EDM damage usually occurs in the outer race. The multiple "ruts" that are formed in the outer race generate high levels of modulating ball pass frequency of the outer race.
4. There are hundreds of on/off switches of the power transistors visible in the voltage traces. This waveform is smoothed by the impedance of the winding system, as seen by the relatively smooth current waveform.
5. Zero.
6. This motor is on a VFD, and the speed is varied during normal operation.
7. Again, most EDM damage occurs in the outer race, so these impacts would be separated by the time interval that would represent ball pass frequency of the outer race.
8. A Rogowski coil, signal conditioner, and 20-MHz oscilloscope.
9. The approximate peak-to-peak voltage measurement is 180 mV, which would equate to more than 3.5 A. Yes, this motor is in trouble.
10. The OEM stripped the outer insulation from the braided ground surround and then clamped this to the back plate with a metal clamp. This is not acceptable, and the OEM measured shaft currents that were over 1 A.
11. 0.5 A.
12. False. To date, there is *not one* motor manufacturer that will warranty its motor against EDM damage, including if it has shaft grounding rings or other shaft-grounding brushes applied. The reason? They don't work.

CHAPTER 9

1. The fault is affecting both power heads equally. This could be a fault in the signal from the controller in the VFD, or it could be the encoder. Replacing the encoder is a simple and very inexpensive item and should be the first step to alleviating the problem.
2. The base of the peak indicated resonance. The most probable answer is that the faulty encoder created torsional impacts in the incoming voltage waveforms that in turn excited the resonant frequency of the motor support frame.
3. The encoder
4. The encoder should be rigidly mounted so there can be no phase difference between the source from the application and the housing of the encoder.
5. These are current readings. There is a serious current drain on I2. I3 is also affected. This is the result of a bad VFD output.
6. 4%.
7. Current readings indicate an unbalance of more than 26%. This is extremely high and requires further investigation. The result was identification of loose connections in the motor terminal box, which were black from arcing. This is a serious level of unbalance and will always require immediate attention.
8. First, a word of caution: The PdMA EMAX tester is only safe to use at voltage levels of 600 V or less. Notice that this motor was operating at 4,600 V. For this test, the motor controls cabinets are usually provided with a voltage transformer. This particular motor cabinet had a 39:1 voltage transformer. This yields voltages in the 120-V range, which is more than safe.

   The results indicate a current imbalance. This is normally due to lead resistance differences. The plant, during the next shutdown, discovered an issue with the power leads at the incoming power transformer and corrected the problem. The next test in 6 months indicated no problem. See the following image.

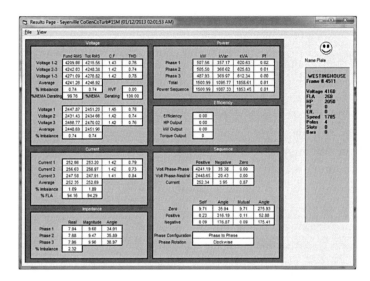

9. The motor that required longer to achieve full speed is also operating at a higher current draw. The fans require a study of their individual loading. Further examination of these fans indicated a turbulence issue that was affecting the faulty fan.

10. Faulty construction, improper application, including frequency starts and stops or excessive starting torques.

11. False. The forces on the bars during normal operation make their identification much easier.

12. False. The amount of current use by most common static testers is not near the recommended level of 70% of nameplate current, and as a result, any amount of material that still connects the bar to the end ring will carry the current. Again, faulty bars in a rotor are best tested at load.

13. Throw it away. It's ruined. It will now warp as it approaches operating temperature. Never swedge the rotor bars in one of these rotors.

14. (1) The current signature will have sidebands on either side of the synchronous peak, and these sidebands will be less than 45 dB below the synchronous peak. (2) The vibration signature will have ± pole pass sidebands on either side of 1×, 2×, and especially 2× line frequency peaks. (3) The third item, which is more difficult to achieve in this day of digital meters, is a more than 5% fluctuation in the current meter.

15. This is the result of a core loss test on an aluminum cast rotor. The current is
induced into the rotor to bring the connection points up close to operating
temperature. Then an infrared camera is used to look for hot spots, or points
carrying excessive current due to insufficient material.

Another test procedure recently proven to work well calls for high
current from a core-loss tester directly connected to the end ring of the
rotor. Magnetic paper is then laid on the rotor body to identify spots
where the current is interrupted.

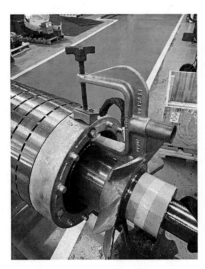 

Index

Note: Entry numbers followed by *f* are for figures; *t* are for tables.

ABOUT THE AUTHOR

RON BROOK resides in Medford, New Jersey. His career in vibration analysis of rotating machinery started in 1975. The job included the analysis of the machine as well as the "hands on" repair, including shaft and bearing changes, dynamic balancing of air handlers, dial indicator alignments, and more. Ron started a family and bought a house while working in the field. That's when he decided to use his G.I. Bill to attend college. Days in the field, nights getting a B.S. degree from Rutgers.

His stints with firms such as Nicolet Scientific, Zonic Corporation, and REM Technologies, all played a part in compiling a complete set of tools to use in the field. In 1993, he joined Reliance Electric at their Philadelphia Service Center. He continued to work out of the same building for the next 22 years through corporation changes, Rockwell Automation, Baldor, and finally Integrated Power Services.